SUSTAINABLE ENERGY AND ENVIRONMENT

An Earth System Approach

SUSTAINABLE ENERGY AND ENVIRONMENT

An Earth System Approach

Edited by

Sandeep Narayan Kundu, PhD
Muhammad Nawaz, PhD

Apple Academic Press Inc. | Apple Academic Press Inc.
3333 Mistwell Crescent | 1265 Goldenrod Circle NE
Oakville, ON L6L 0A2 | Palm Bay, Florida 32905
Canada USA | USA

First issued in paperback 2021

Exclusive worldwide distribution by CRC Press, a member of Taylor & Francis Group

No claim to original U.S. Government works

ISBN 13: 978-1-77463-427-1 (pbk)
ISBN 13: 978-1-77188-763-2 (hbk)

Library and Archives Canada Cataloguing in Publication

Title: Sustainable energy and environment : an earth system approach / edited by Sandeep Narayan Kundu, Muhammad Nawaz.

Names: Kundu, Sandeep Narayan, 1974- editor. | Nawaz, Muhammad, 1966- editor.

Description: Includes bibliographical references and index.

Identifiers: Canadiana (print) 20190156775 | Canadiana (ebook) 20190157658 | ISBN 9781771887632 (hardcover) | ISBN 9780429430107 (ebook)

Subjects: LCSH: Renewable energy sources—Environmental aspects.

Classification: LCC TJ808 .S87 2020 | DDC 333.79/4—dc23

CIP data on file with US Library of Congress

Apple Academic Press also publishes its books in a variety of electronic formats. Some content that appears in print may not be available in electronic format. For information about Apple Academic Press products, visit our website at **www.appleacademicpress.com** and the CRC Press website at **www.crcpress.com**

About the Editors

Sandeep Narayan Kundu, PhD

Sandeep Narayan Kundu, PhD, is a geoscientist with over 20 years of experience in industry and academics alike. He currently serves as an adjunct faculty at the Department of Civil and Environmental Engineering, National University of Singapore (NUS), and also works full time for Fugro Singapore Marine as a geoscientist. Dr. Kundu worked in various capacities in the energy exploration industry in the geoscience domain at multinational organizations such as BHP Billiton Fugro Surveys and Reliance Industries. He also did research at Friedrich Schiller University, Jena (Germany); the Institute of Geoinformatics, Muenster (Germany); the Indian Institute of Technology (ISM, India); and the Indian Institute of Technology, Kanpur (India). Dr. Kundu is a recipient of the prestigious British Council Fellowship and the Junior Research Fellowship in Earth Sciences. He has delivered keynote addresses at international forums, which include the likes of the Institute of Indian Geographers, IQPC, and PRAXIS conferences. As a SEAPEX Senior Lecturer at NUS, he designed and developed the Petroleum Geoscience and Petroleum Exploration educational modules, which he taught at the Faculty of Engineering and the Faculty of Arts and Social Sciences. His primary research interests involve geospatial technology applications for exploration of natural resources, map the seafloor, and evaluate environmental change.

Muhammad Nawaz, PhD

Muhammad Nawaz, PhD, is geomorphologist with expertise in geographic information systems (GIS), decision support systems, and environment science. Since 1993, Dr. Nawaz has served at several different universities including Punjab University in Pakistan and the Faculty of Geoinformation Science and Earth Observation (ITC), the University of Twente in the Netherlands. He has vast experience and knowledge in design, development, delivery, coordination, and teaching of courses related to physical geography, GIS, and environmental science to undergraduate, graduate, and postgraduate students. Dr. Nawaz has over 20 years of research experience in a broad range of issues relating to catchment management and sustainable development. Before joining the National University of Singapore, Dr. Nawaz worked as a Lecturer of Environment Science eLearning at Charles Darwin University, Australia, and before this he has worked as Associate Head/Assistant Professor at the GIS Centre, University of the Punjab, Pakistan. He has conducted multidisciplinary research in several catchments including Darwin Harbour and Daly River catchment, Australia; Mangla Reservoir catchment, Pakistan; and Saddang Watershed, Tana Toraja, South Sulawesi, Indonesia. Dr. Nawaz's expertise includes sediment source identification, multicriteria evaluation, and GIS applications in decision-making. Dr. Nawaz is a NFP (Netherlands Fellowship Programme) fellow and currently works as Faculty at the Department of Geography, of National University of Singapore where he teaches subjects that include biophysical environment, planet earth, energy resources, coastal environments, and spatial data handling.

Contents

Contributors

Madhumita Das, PhD
Fakir Mohan University, Balasore, Odisha, India

Deepak Dash, MSc Tech
Reliance Industries Limited, Thane-Belapur Road, Navi Mumbai 400701, Maharashtra, India

Shreerup Goswami, PhD, DSc
Department of Earth Sciences, Sambalpur University, Burla 768019, Odisha, India

Ratiranjan Jena, MSc
Faculty of Engineering and Information Technology, School of Systems, Management, and Leadership, University of Technology Sydney, Ultimo, Australia

Munukutla Sruti Keerti, MS
SoUL Program, Indian Institute of Technology Bombay, Mumbai, India

Sandeep Narayan Kundu, PhD
Department of Civil & Environmental Engineering, National University of Singapore, Singapore

Nausheen Mazhar, MSc
Department of Geography, Lahore College for Women University, Lahore, Pakistan

Debadutta Mohanty, PhD
CSIR-Central Institute of Mining and Fuel Research, Dhanbad 826015, India

Muhammad Nawaz, PhD
Department of Geography, National University of Singapore, Singapore 117570, Singapore

Satyabrata Nayak, PhD
Petronas Twin Towers, Kuala Lumpur, Malaysia

Lokanath Peddinti, MTech
Reliance Industries Limited, Thane-Belapur Road, Navi Mumbai 400701, Maharashtra, India

Binapani Pradhan, MSc
Department of Environmental Science, Utkal University, Bhubaneswar, Odisha, India

Biswajeet Pradhan, PhD
Faculty of Engineering and Information Technology, School of Systems, Management, and Leadership, University of Technology Sydney, Ultimo, Australia

Sanghamitra Pradhan, MSc, MPhil
Department of Earth Sciences, Sambalpur University, Burla 768019, Odisha, India

Sanjib Kumar Sahoo, PhD
Department of Electrical and Computer Engineering, National University of Singapore, Singapore 117570, Singapore

Abani Ranjan Samal, PhD
GeoGlobal, LLC, Salt Lake City, Utah, USA

Maher Ibrahim Sameen, MSc
Faculty of Engineering and Information Technology, School of Systems, Management, and Leadership, University of Technology Sydney, Ultimo, Australia

Farha Sattar, PhD
School of Education, Charles Darwin University, Australia

Loo Mei Yee, BSc
Department of Geography, National University of Singapore, Singapore 117570, Singapore

Sahar Zia, MPhil
Department of Geography, Lahore College for Women University, Lahore, Pakistan

Abbreviations

2D	two-dimensional
3D	three-dimensional
AAPG	American Association of Petroleum Geologists
AHP	analytical hierarchy process
AMD	acid mine drainage
ANN	artificial neural networks
ARI	Advanced Resources International
BSR	bottom simulating reflectors
BUWT	building mounted urban wind turbines
CAPP	Canadian Association of Petroleum Producers
CBM	coalbed methane
CCS	capture and sequestration
CMM	coal mine methane
CSG	coal seam gas
CSP	concentrating solar power
DEM	digital elevation model
ED	electrodialysis
EDR	electrodialysis reversal
EF	evolutionary fauna
EIA	Energy Information Administration
EMF	electric and magnetic fields
EPA	Environmental Protection Agency
FL	fuzzy logic
FR	frequency ratio
GDP	gross domestic product
GHG	greenhouse gas
GHSZ	gas hydrate stability zone
GIS	geographic information system
GPS	global positioning systems
GTS	geological time scale
HAWT	horizontal-axis wind turbines
HI	hydrogen index
IDS	inverse distance square

InSAR	SAR interferometry
ISL	in situ leaches
ISR	in situ recovery
LCOE	levelized cost of electricity
LiDAR	light detection and ranging
LoM	life of mine
LOME	late Ordovician mass extinction
LR	logistic regression
Mb	body–wave magnitude scale
MEMS	microelectromechanical systems
$MeNH_2$	monodentate methylamine
MIR	mid-band infrared
ML	local magnitude scale
MORB	mid-oceanic-ridge basalt
Ms	surface–wave magnitude scale
MSW	municipal solid waste
MW	megawatts
Mw	moment magnitude scale
NFP	Netherlands Fellowship Programme
NIR	near infrared
NORM	naturally occurring radioactive material
NO_x	nitrogen oxides
NPV	net present value
NSIDC	National Snow and Ice Data Center
NUS	National University of Singapore
OIB	ocean island basalt
OOIP	original oil in-place
OTEC	ocean thermal energy conversion
PV	photovoltaic
RCPs	representative concentration pathways
SAGD	steam-assisted gravity drainage
SAR	synthetic aperture radar
SEAPEX	South East Asian Petroleum Exploration Society
SPE	Society of Petroleum Engineers
SPEE	Society of Petroleum Evaluation Engineers
SVM	support vector machine
TDS	total dissolved solids
TGS	tight gas sand
TIR	thermal infrared

TLS	terrestrial laser scanning
TOC	total organic carbon
TPES	total primary energy supply
TR	transformation ratio
UAV	unmanned aerial vehicles
UBD	underbalanced drilling
VAWT	vertical-axis wind turbines
WBM	water-based mud
WLC	weighted linear combination
WPC	World Petroleum Congress

Preface

The study of earth, energy, and environment is critical to understanding where life stands amid all the growth and development around us. Understanding how Earth has evolved and how its subsystems interact and influence various energy resources and impact the environment we live in is critical to sustaining life on the planet.

The way we study and think about Earth is rapidly changing, especially in the wake of climate change and global warming. From observing local climate and weather we are moving toward understanding the processes on a global scale. In addition, we have identified a strong human influence on our environment through increased use of energy in our lives. This has influenced the natural cycles of interactions between the different subsystems of Earth. Rapid advances over the past couple of decades in the way we understand Earth, energy, and our environment make it essential to have continual reappraisals of teaching methods and the contents of the textbooks. The modern approaches to understanding the nexus between Earth, energy, and environment have, in general, not yet found their way beyond scientific papers into textbooks.

Various courses on Earth, energy, or environment are being taught across academic departments around the globe for preparing learners for specific careers. Today, the focus is on interdisciplinary holistic education that leverages basic sciences to achieve a strong applied sense in individuals. We have seen many books that depict environmental science with the perspective of fossil fuels and renewable energy sources. We also come across books that speak about earth system science. But despite having many academic and governmental agencies that use Earth, energy, and environment as a name, a fundamental academic book involving the three in an integrated manner is lacking.

Our aim is to help to fill this gap, particularly for those studying introductory earth processes, energy fundamentals, and introductory environment courses. At the same time students in related disciplines such as ecology, agriculture, petroleum, and gas and indeed all those are interested in earth, environment, and energy should find a basic introduction to modern ideas in these fields in this book. Some of the concepts that are introduced undoubtedly go rather beyond the general scope of basic level courses, so that the book will also serve as the foundation for more advanced studies. It is hoped that this book will give the reader sufficient appreciation of earth, energy, and environment interactions to apply these ideas elsewhere himself.

The book aims to provide a holistic understanding of earth, energy, and environment through careful selection of invited chapters. The chapters elaborate the interactions between the different parts of earth; how energy is exchanged between the atmosphere, hydrosphere, biosphere, and geosphere and impacts the environment we live in. It extends earth system science to the energy and environment perspective with an aim to emphasize the rationale for balancing the forces for a sustainable future. This book, through its treatise on the fundamental science of earth, energy, and environment, deals with the complex interactions that influence the way our earth systems work. This book introduces the developing science of Earth, energy, and environment to the students.

The book begins with an introduction to Earth, energy, and environment that sets the scene for the following three sections on Earth, energy, and environment, respectively. The introduction highlights the strong nexus between Earth, energy, and environment through connectivity and interactivity of natural processes which influence the sustainability of life. The first section deals with Earth, its origin and evolution, its surface and rock-forming processes, with strong reference to past climates, climate changes, and influencing planetary cycles. Four chapters in this section deal with each of these aspects of Earth. The next section is on energy, its fundamentals and its resources—conventional and unconventional fossil fuels and renewable energy sources and aspects of energy availability and economics. The fact that energy is central to life and its resources is critical to activity is elaborated here. This section brings forth the connection between different energies formed with Earth's surface and subsurface processes. The final section discusses the environment, its impact on human health and well-being, climate change and resulting environmental impacts, environmental disasters from exploration, production, and usage of energy and balancing energy usage and its environmental impacts.

Although a holistic approach to the integrated science of earth, energy, and environment has been carefully considered in the organization of the book, each chapter in itself can stand in entirety, without losing the continuity of the flow of information from the previous ones. Special attention is devoted to the photographs and illustrations, making them region agnostic, which provides an opportunity for leaders to relate them to their own surroundings.

—**Sandeep Narayan Kundu**
Muhammad Nawaz

PART I
EARTH

"Scientists still do not appear to understand sufficiently that all earth sciences must contribute evidence toward unveiling the state of our planet in earlier times, and that the truth of the matter can only be reached by combing all this evidence. ... It is only by combing the information furnished by all the earth sciences that we can hope to determine 'truth' here, that is to say, to find the picture that sets out all the known facts in the best arrangement and that therefore has the highest degree of probability. Further, we have to be prepared always for the possibility that each new discovery, no matter what science furnishes it, may modify the conclusions we draw."

—Alfred Wegener

CHAPTER 1

Weathering, Erosion, and Deposition

FARHA SATTAR[1,*], MUHAMMAD NAWAZ[2], and
SANDEEP NARAYAN KUNDU[3]

¹College of Education, Charles Darwin University, Darwin, Australia

²Department of Geography, National University of Singapore, Singapore

*³Department of Civil & Environmental Engineering,
National University of Singapore, Singapore*

*Corresponding author. E-mail: farha.sattar@cdu.edu.au

ABSTRACT

Weathering, erosion, and deposition are the surface processes that shape the topography of the earth. Weathering is the breaking down of the upper thin layer of the solid crust by physical and chemical means. Erosion is the process of transportation of the weathered rock matter. The transported sediments accumulate in suitable environments where they settle down forming sedimentary deposits. This chapter covers the weathering processes, and describes the erosional and depositional processes, and erosional and depositional environments.

1.1 INTRODUCTION

Weathering, erosion, and deposition are sequential processes which are driven by the interaction and material exchanges between lithosphere, hydrosphere, and atmosphere interaction. It is the most common surface process which is ongoing since Cambrian age. Weathering, erosion, and deposition produce sedimentary rocks which were absent during the pre-Cambrian era as the hydrosphere and atmosphere were not as evolved at that stage than at the present date.

Weathering is the breaking down of the upper veneer of the solid crust by physical and chemical means. Existing rocks are broken down into smaller pieces as a result of climatic variation at the diurnal, seasonal, and decadal scale. Water and its phase changes due to temperature variation cause weathering. Water being a solvent takes many salts in the rocks into solution. Rainwater is acidic in nature, as it imbibes carbon dioxide and nitrous and sulfur compounds in the atmosphere, and can alter certain minerals in the rocks and take the resultant anions and cations into the liquid phase.

Erosion is the process of transportation of the weathered rock matter. Once rocks and minerals are broken down, they are removed from their location of production to other location by wind and flowing water. Water in rivers and streams can carry the weathered rock matter in solution, suspension, saltation, and rolling modes whereas in deserts wind can blow sand and smaller particulates. Weather material can also travel downslope due to the effect of gravity forming rock fall, landslides, mudslides, and slumps. In lesser steep regions, weathered topsoil can move slowly undetected and is called a creep.

The transported sediments accumulate in suitable environments where they settle down forming sedimentary deposits. This happens at later stages of a river where it is heavily laden with sediments and meanders slowly in the plans shedding its sediment load at meanders and channels. At deltas, where the river meets the seas, the waves and tides hind the pace of the river and push back the sediments along the beach forming beach deposits. If the river drains into a lake its water velocity is drastically impacted which leads to the settling of the suspended sediments. Increased concentration of salts in lakes leads to their precipitation forming chemical deposits. In the following sections, the processes of weathering, erosion, and deposition are elaborated along with the resulting deposits and landscapes by various agents in different depositional settings.

1.2 WEATHERING PROCESSES

The term weathering generally refers to the processes that break down the materials either at the surface or near the surface of the earth (Dixon, 2004). Gupta (2011) expands this definition to include how weathering involves interactions between the crustal material, atmosphere, biosphere, and hydrosphere. Allen (1997) describes the weathering as the process, as being in situ, whereby weathering occurs on the rock itself. Bringing these definitions together, we will use the following working definition of weathering for this chapter as: the weathering is a group of processes that are the result of interactions between crustal material and other earth systems, to cause a physical breakdown and/or chemical alteration of crustal material in situ and form more stable products.

The role of weathering is key to understanding soil formation processes and landform evolution. It is also closely embedded in the various sediment recycling systems. Thus, this section outlines the forms and processes in which crustal materials are weathered to produce sediments, as well as factors which may affect the rate of weathering.

First, it is examined how rocks can be mechanically broken into smaller pieces. As a result of physical weathering, rocks would break down into smaller pieces in four main forms. They may be classified broadly into blocky (Fig. 1.1) or granular disintegration. Granular disintegration (Fig. 1.2) refers to the physical disintegration of rock into individual grains and rock crystals (Nicholson, 2004). Instead of large rock pieces, granular disintegration forms loose debris that may be coarse or fine-grained. These grains are then eroded by water or wind action to reveal a fresh section of rock for weathering. This usually results in a pitted rock surface.

FIGURE 1.1   Blocky disintegration.

FIGURE 1.2   Granular disintegration.

Block disintegration refers to the breakdown of rock into smaller fragments. They include exfoliated horizontal blocks from dilation and the shattered fragments from freeze-thaw weathering. It can take a cuboidal or fragmentary appearance. Small-scale exfoliation, whereby the outer layer of rocks peels away in small platy pieces, causes spheroidal weathering, thus resulting in a rounded cornerstone.

1.2.1   PHYSICAL WEATHERING PROCESSES

Physical weathering describes the disintegration of various rocks into smaller pieces. In this process, the properties of the new pieces are identical to the original rock.

1.2.1.1   SALT WEATHERING

Salt weathering, also known as haloclasty, occurs in rock pores and cracks. It involves the precipitation and growth of salt crystals from salt solutions or the volumetric expansion of salts during hydration. Salt crystal growth and expansion exert pressure along the pores and cracks that cause the disintegration when the force exceeds the tensile strength of the rock (Evans, 1970). The expulsion of mineral grains also results in granular disintegration and rock pore enlargement to form a honeycombed network of pits on the surface (Fig. 1.3).

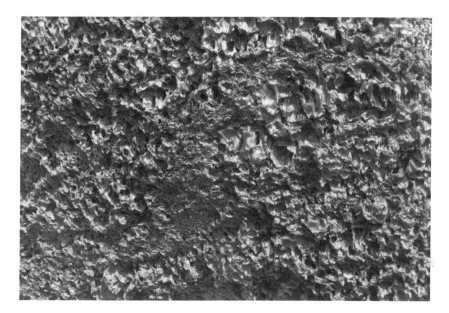

FIGURE 1.3   Honeycomb patterns due to salt weathering.

Some salts, such as $NaNO_3$ and $MgSO_4$, precipitate because their solubility decreases as temperature falls at night. Nocturnal cooling causes salts to crystallize, producing large volumes of crystals that push against the wall of the rocks. Repeated cycles of pressure exertion would slowly cause the rock to break apart along these cracks.

Other salts, such as $CaSO_4$ and Na_2CO_3 are prone to hydration and dehydration with reference to changes in the temperature and the water. As water is absorbed, salt volume increases and exerts pressure up to 100 MPa against the pore walls (Trudgull, 2004). Research by Winkler and Wilhelm (1970) has shown that the greatest hydration pressure occurs when anhydrite hydrates into gypsum, with a pressure far exceeding the tensile strength of the rock.

1.2.1.2   FROST WEDGING

Frost wedging breakdowns rock through alternate freezing and thawing cycles of water in the pores and cracks of the rocks (Fig. 1.4). It is very common in temperate or mountainous climates, where diurnal temperature ranges above and below the freezing point of water. Two models that are commonly used to describe this process are: the volumetric expansion model and the segregation ice model.

FIGURE 1.4   Enlargement of joints by frost weathering.

Liquid water can expand up to 9% as it freezes into ice. Water in a fully filled rock joint would theoretically generate pressures up to 207 MPa as it freezes at −22°C (Tsytovich, 1975). The ice-induced force fractures the rock and this constitutes the volumetric expansion model. The ice-segregation model states that during freezing, the ice nuclei attract liquid water molecules from the adjoining pores and capillaries due to a difference in chemical potential gradient. As the water migrates toward these sites, it creates microcracks and fissures in the rock (Hallet et al., 1991). Repeated microfracturing would cause the rock to break apart eventually.

1.2.1.3   INSOLATION WEATHERING

Insolation weathering takes place in areas with large diurnal temperature ranges, such as hot desert environments. Daytime temperatures may exceed 40°C, while night-time temperatures can drop to 9–10°C. Similarly, the temperature of the outer layer of the rock also fluctuates greatly, reaching up to 80°C in the day (Goudie, 2004). However, as rocks are not good conductors of heat, therefore the internal temperature of the inner rock layers does not vary to the same magnitude. This difference in the rate of heating and the existence of a temperature gradient within the rock means that the outer layer experiences larger deformations than the inner layer. This creates high internal tensile stresses which cause the rock to bend and buckle (Fig. 1.5).

FIGURE 1.5   Process of insolation weathering leading to exfoliation.

After continual cycles of heating and cooling, the bonds between the outer layer and inner layers of rock would weaken, causing the outer rock layers to gradually peel off through the process of exfoliation.

1.2.1.4   PRESSURE RELEASE

Plutonic rocks such as granite and gabbro are buried under a deep cover of rock overburden which exerts lithostratigraphic stress. This compressive stress is balanced by an opposing internal resistive stress. Denudation of the overburden exposed the plutonic rocks to an environment of lower overlying pressures resulting in the expansion of the rocks (Goudie, 2004), resulting in jointing. Joints are planar cracks in plutonic rocks that are exposed to the surface examples of which are columnar basalts. Ingression of rainwater in the joints aids the separation of the rock into blocks and other forms of weathering result in the separation.

1.2.1.5   CHEMICAL WEATHERING

Chemical weathering is the disintegration of rocks via chemical processes on their minerals (Summerfield, 1991). These processes typically occur at the rock surface and are driven mainly by water and oxygen under acidic conditions to form new minerals from the primary minerals originally present in the rocks. There are four main types of chemical weathering: Carbonation, Hydrolysis, Solution, and Hydration.

1.2.1.6   CARBONATION

Carbonation is a reaction that involves CO_2 and water with carbonate minerals such as calcite ($CaCO_3$). CO_2 first dissolves in rainwater or water vapor to produce H_2CO_3 (aq), as seen in eq 1.1.

$$CO_2 \, (g) + H_2O \, (l) \rightarrow H_2CO_3 \, (aq) \qquad (1.1)$$

Carbon Dioxide + Water → Carbonic Acid

The weak carbonic acid then reacts with calcareous minerals, dissolving them into solution and thus weathering away the rock. Soluble minerals are flushed away by water and accumulated at the groundwater reserve. $CaCO_3$, a common mineral in limestone, is undergoing carbonation via eq 1.2 (Huggett, 2011):

$$CaCO_3(s) + H_2CO_3 \, (aq) \rightarrow Ca^{2+} \, (aq) + 2HCO_3^- (aq) \qquad (1.2)$$

Calcite + Carbonic Acid → Calcium + Hydrogen Carbonate

The cementing $CaCO_3$ that holds rocks together dissolves, eroding the limestone to form caves. Huge cavities are formed by the progressive movement of mildly acidic water flowing to the bottom of the rock.

1.2.1.7 HYDROLYSIS

Hydrolysis is the chemical breakdown of a substance when combined with water. When minerals are hydrolyzed, clay minerals and crystalline rocks, such as silica, are formed (Dixon, 2004). The most common mineral that undergoes hydrolysis is feldspar found in granite. Rainwater percolates through the overlying crustal material and reacts with granite. Feldspar then reacts with water to form clay minerals that weaken the rock. Equation 1.3 shows orthoclase, an alkali feldspar, with water under tropical acidic conditions to form kaolinite, silicic acid, and potassium ion (Gupta, 2011):

$$2KAlSi_3O_8 (s) + 2H^+ (aq) + 2H_2O (l) \rightarrow Al_2Si_2O_5(OH)_4 (s)$$
$$+ 4SiO_2 (s) + 2K^+ (aq) \tag{1.3}$$

K-feldspar + Water (acidic) → Kaolinite + Silica + Potassium

Water acts as a weak acid on silicate minerals, dissolving them to release cations and silica. This reaction is important because the substances formed in these reactions are essential for plant growth, such as the positively charged metal ion K^+ and soluble silica. The resulting kaolinite remains in situ and until it is eroded.

Deep weathering of tropical granite occurs as water percolates underground to weathering profile. It consists mainly of regolith, which is an "incoherent mass of varying thickness composed of materials essentially the same as the makeup of the parent rocks themselves, but greatly varying conditions of mechanical aggregation and chemical combination" (Merrill, 1897). The bottom layer consists of fresh bedrock. Above it, unweathered cornerstones are buried in a matrix of weathered kaolinite. Saprolite, a grey rock that retains the form of a cornerstone yet has been entirely chemically altered is also found here. The top is covered in a heavily weathered clay-rich horizon capped by organic soils.

1.2.1.8 HYDRATION

Hydration is the process whereby entire water molecules and minerals form chemical bonds to create a different molecular structure. While in theory, this process is similar to salt weathering (see p 3), but the difference in chemical hydration is the formation of new more complex layered minerals that expand and disintegrates by becoming soft and losing their luster. Hydration aids other chemical reactions by "introducing water molecules deep into crystal structures" (Summerfield, 1991: 133). Minerals formed through weathering, such as aluminum oxide, iron oxide, and calcium sulfate, undergo hydration.

1. Hematite [Iron (III) Oxide] forming bright reddish-brown limonite

$$2Fe_2O_3 + 3H_2O \rightarrow 2Fe_2O_3.3H_2O$$

Hematite (red) + Water → Limonite (yellow)

2. Bauxite [Aluminum (III) Oxide] forming hydrated aluminum oxide

$$Al_2O_3 + 3H_2O \rightarrow Al_2O_3 .3H_2O$$

Bauxite + Water → Hydrated Aluminum Oxide

3. Anhydrite [Calcium (II) Sulfate] forming gypsum

$$CaSO_4 + 2H_2O \rightarrow CaSO_4 .2H_2O$$

Anhydrite + Water → Gypsum

The converse can happen, where the mineral loses water to become anhydrous in a dehydration reaction. For example, limonite may undergo dehydration to form hematite.

$$2Fe_2O_3 .3H_2O \rightarrow 2Fe_2O_3 + 3H_2O$$

Limonite → Hematite + Water

1.2.1.9 REDOX REACTIONS

Oxidation is the reaction of rock minerals with oxygen at the Earth's surface. The most common minerals that undergo oxidation are those bearing iron. When iron reacts with oxygen, it loses an electron to form reddish-brown iron oxide. As iron oxide is more fragile and less malleable than iron, it is weakened and crumbles easily upon oxidation, allowing the rock to be more susceptible to erosion. As iron can have several oxidation states (Fe^{2+}, Fe^{3+}, etc.), it gives rise to several possible products as well.

One example is the oxidation of Magnetite to give Hematite (Summerfield, 1991).

$$4Fe_3O_4(s) + O_2(g) \rightarrow 6Fe_2O_3(s)$$
Magnetite + Oxygen $\rightarrow$ Hematite

Hematite is brownish-red in color, giving rise to the reddish color of tropical weathered rocks. Hydration can occur in conjunction with oxidation to give rise to limonite (as mentioned above), which has a yellow color. On the other hand, the reverse process of removing oxygen and gaining an electron is called reduction. This occurs under conditions with excess water and minimal oxygen levels.

$$2Fe_2O_3(s) - O_2(g) \rightarrow 4FeO (s)$$
Hematite + Oxygen $\rightarrow$ Ferrous Oxide (Reduced Form)

1.2.2 BIOLOGICAL WEATHERING PROCESSES

Biological weathering is the breakdown of rock through processes which are organic in origin. This includes tree root penetrative growth into existing rock cracks, animal weathering where their decomposition of organic materials can lead to the production of organic acids with weathering capability, as well as lichen and microorganisms contributing their own weathering effects when situated on a rock as their substrate. It is important to note that while the three categories of physical, chemical, and biological weathering are widely used in the geography field, often these processes are involved with each other synergistically. While physical and chemical processes may represent a larger portion of total rock weathering instances, this weathering hardly occurs in a sterile environment, meaning that the weathering will actually be defined through both biophysical and biochemical processes in nature.

1.2.2.1 ROOT PENETRATIVE GROWTH

Root growth exerts pressures along crevices or voids in rocks, causing the rock to crack or break. Tree root wedging is a biophysical weathering process, but tree roots are also capable of biochemical processes: The microbial communities that colonize young tree roots are also capable of rock weathering. A 2004 study indicates that these microbial communities can weather igneous, metamorphic, and sedimentary rocks (Puente et al., 2004). Besides the decomposition of organic material, tree roots also facilitate the production of carbonic acid through the dissolution of respirated CO_2 with water. These acids then react with $CaCO_3$ and $MgCO_3$ to form calcium bicarbonates through carbonation.

1.2.2.2   EFFECTS OF LICHEN

Lichen also causes both biophysical and biochemical weathering. Lichens are organisms formed out of symbiotic fungi and alga. They are found on surfaces that experience good air quality, sunlight, water, and nutrients, such as rock surfaces. Lichens can store water for long periods of time, thus ensuring the constant presence of water for chemical weathering. In addition, as lichens expand and contract by their water content changes, it creates microfractures in the rock substrate (Moses and Smith, 1993; Fry, 1927).

As noted by Robinson and Williams (2000), the distribution of lichen rock weathering is extremely complex, depending on microenvironmental conditions. Variables such as rock slope, air quality, slope aspect, and photosynthesis rates in the algal part of lichen growths all have influence over the effective rock weathering of the lichen's substantial rock. Danin and Garty's landmark study (1983) indicates that endolithic lichens in the Negev Highlands of Israel are responsible for jigsaw puzzle like burrowing in rocks.

Wilson and Jones (1983) determined that lichen weathering occurs when the lichen's dominant fungal component excretes oxalic acid, which reacts with $CaCO_3$ in the substrate rock to form calcium oxalate. Oxalic acid contains large amounts of protons and is highly water-soluble (Dixon, 2004) forming a strong weathering solution that donates protons for chemical weathering processes (Barker et al., 1997). As such, lichen weathering should be characterized as a type of biochemical weathering. Within this category, there are different visual indications of weathering that include mineral etching (Fig. 1.6), and secondary weathering products like silica gel and calcium oxalate.

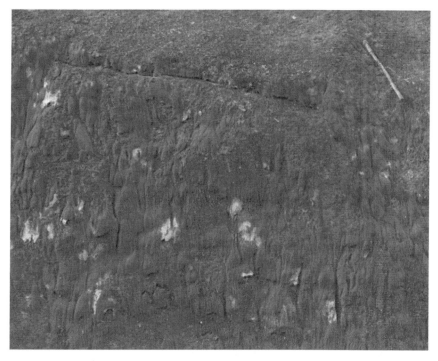

FIGURE 1.6 Lichen action produces mineral etching on rock surface.

1.2.2.3 CHELATION

The partial decomposition of organic material produces organic compounds which can help in breaking down the materials. These include humic, fluvic, citric, malic, and gluconic acids (Viles, 2004) which can form chelates. Chelates are compounds containing ligands which act as "claws" to extract metal cations from minerals found in the matrix to form a ring-like structure encompassing the metallic ion. Chelates are defined as a compound that contains a ligand, a typical organic compound, chemically bonded to a central atom of the metal at two or more than two points (Trudgull, 2004). A ligand then is defined as "an ion or molecule attached to a metal atom by coordinate bonding" (Trudgill, 2004).

These coordinate bonds, which can be covalent or ionic in nature, usually occur in even pairs. There can be 2, 4, or 6 coordinate bonds surrounding the metal cation in an octahedral arrangement. The carboxylic and/or amine groups present in the chelate helps electrons to form the coordinate bonds. The efficacy of chelation depends on the denticity. For example, two molecules of monodentate methylamine (MeNH$_2$) are needed to full a complex Cu^{2+} ion, but only a single molecule of bidentate ethylenediamine (en) is needed. As a result, it is thermodynamically more efficient to form a [Cu(en)]$^{2+}$ complex than a [Cu(MeNH2)$_2$]$^{2+}$ complex due to less number of molecules involved in the ethylenediamine process and resulting lower entropy changes.

Chelation most commonly affects iron and aluminum ions (Turdgill, 2004). However, zinc, copper, manganese, calcium, and magnesium are also possible chelation candidates depending on the type of chelating agent present. As chelation weathering extracts these metal ions through the decay of organic matter, some argue that chelation is a more significant mineral weathering process as compared to hydrolysis (Huang and Keller, 1972). This is significant especially in the weathering of silicate-rich igneous rocks in the tropics. The resulting weathered products include dissolved or amorphous silica, which can be removed by water. Chelation also produces hydrogen ions in the process, which facilitates further hydrolysis reactions as described above.

1.2.2.4   HUMAN ACTIVITIES AND WEATHERING

Construction works are one of the activities that increase the rate of weathering. Rock is being broken down and cut into to make way for buildings, pavements, and dams. This not only causes disintegration of the rock but also increases the surface area of the rock, facilitating other modes of weathering.

Mining the different minerals and metals available on Earth's surface is mainly done via explosives. This results in large amounts of rock being disintegrated at one time, causing the structure of the rock to lose its integrity over time and be easily acted upon via other agents of weathering.

1.2.3   FACTORS AFFECTING WEATHERING

Rock structure, climate, and vegetation are important factors that affect the rock weathering. These factors are described in the following section.

1.2.3.1   ROCK STRUCTURE

Most rocks are not uniform all through, they differ in mineralogy, rock jointing, permeability, porosity, tensility, rock strength, and so on. These differences play their role in rock weathering as described below.

1.2.3.1.1   Mineralogy

Minerals have different chemical weathering rates. Mafic-silicates such as pyroxene and olivine weather faster than silica-rich quartz and feldspar. Goldich (1938) attributes this difference to the arrangement of Si–O–Si bonds in the mineral structure. In olivine, isolated SiO_4 tetrahedra ionically bonded to metal cations are more easily weathered compared to the three-dimensional framework of quartz, where each oxygen atom is covalently bonded to two other silicon atoms and Si–O–Si bonds proliferate the entire structure.

Different colored minerals also expand at different rates thus affecting insolation weathering. In granite, dark-colored hornblende would expand more than light-colored quartz minerals, resulting in a matrix of different expansion extent. This exerts strain in the rock structure and makes it susceptible to further weathering.

1.2.3.2   ROCK JOINTING, PERMEABILITY, AND POROSITY

The surface area to volume ratio affects how much of the rock is exposed to weathering, which are affected by rock joint density and rock porosity. Massive granitic plutons with low joint densities have fewer lines of weaknesses for water to enter and hydrolyze the felsic minerals. However, areas of higher joint densities allow tropical deep weathering as water permeates through and further weathers the underlying plutonic rock. Sedimentary rocks also have bedding planes that can be infiltrated easily by water. Tree roots also exploit and grow in rock joints, weakening the structure of the rock by prying it apart. In a similar fashion, rocks with high joint densities also create more sites for freeze-thaw and salt weathering to take place (Sperling and Cooke, 1985) while introducing sites where stress concentrations can take place.

Additional surface imperfections formed by granular disintegration of individual minerals also increase the surface area for weathering to occur. For example, the removal of kaolinite after the hydrolysis of granite reveals a brown pitted surface of residual silica. Depending on where the rock is located, this will promote weathering through freeze-thaw or salt weathering.

1.2.3.3   ROCK TENSILE/COMPRESSIVE YIELD STRENGTH

Rocks have an inherent yield strength which reflects its resistance to breakage (Nicholson, 2004). Rocks with low yield strength are not able to deform to accommodate the pressures exerted by volumetric changes of the rock mass (Hall, 1999), or of the voids and fissures within it. Thus, they would easily break and shatter into smaller pieces. However, as rocks are being weathered, continued cycles of stresses would also slowly cause changes to the extent of jointing, porosity, and mineral cohesiveness. This would reduce the yield strength of the rock and make it easier to weather—one experiment suggests that the strength of the rock on fracture after undergoing cyclic stress is only 75% as compared to the fracture strength of rocks not under cyclic stress (Griggs, 1936).

1.2.3.4   CLIMATE

Precipitation and temperature are two important controls of weathering rates. Water is a key component of many chemical and physical weathering processes, while temperature influences the rate of chemical reactions and the tensile stresses experienced by the rock.

The rate of chemical weathering is highest in hot and wet climates. As described by the Van't Hoff, the rate of the chemical reactions increased by 2–3 times for every 10°C increase in temperature. High diurnal temperature range encourages insolation weathering, while a narrow diurnal temperature range hovering around the freezing point of water facilitates freeze-thaw weathering.

The presence of water is related to precipitation patterns and is essential in driving various chemical weathering processes. Rainwater may contain nitric or sulfuric acids, and aid the formation of carbonic acids for carbonation. The acidic environment also promotes hydrolysis and redox reactions.

The influence of climate on weathering can be seen in Strakhov's (1967) and Peltier's model (1950). According to Strakhov's model, intense chemical weathering dominates in humid tropical climates due to the availability of water and high temperatures to produce great depths of kaolinite and illite clay minerals. Peltier's model agrees with Strakhov but extends his model to include mechanical weathering processes as well. Peltier suggests that high rates of mechanical weathering actually occur in periglacial and boreal climates as he assumes that freeze-thaw action is the most dominant physical weathering process.

1.2.3.5 VEGETATION

Presence of vegetation will determine the rate of biomechanical and biochemical weathering. Tropical rainforests produce about $0.4-1.3 \times 10^6$ kg km^{-2} yr^{-1} of organic matter (Summerfield, 1991), whose decomposition produces various organic acids and chelating agents for biochemical weathering.

Rocks enveloped with soil are susceptible to chemical reactions with water much more than rocks that are not covered by soil, as soil retains rainwater. The variety of organisms, vegetation, and bacteria that resides in the soil also creates an acidic environment for chemical weathering to occur.

1.3 EROSIONAL AND DEPOSITIONAL PROCESSES

Sediments are everywhere. It is intertwined with weathering and erosion and can be transported and deposited through various means. Understanding these processes is vital for us to best navigate our surrounding environment. The following section discusses the finer aspects of sediment transportation and deposition.

Sediment is made out of particles that can be transported by fluid flow such as water and wind. It can be composed of soil-based organic matter or inorganic materials such as rocks and sand, and arises from erosion, weathering, and even from decomposing materials such as algae (Fondriest Environmental, 2014). Sediment can be divided into two categories: bedded or suspended. Suspended sediments are particles floating and/or moving in the water column while bedded sediment describes sediments that have settled to the bottom of the flow channel, for example, estuaries and riverbeds.

It should also be noted that suspended sediments and suspended solids differ only by the method of measurement. Suspended sediment concentration is measured in mg/L by filtration and drying the entire water sample (Fondriest Environmental, 2014). Total suspended solids obtained via subsampling are also measured in mg/L but often excludes larger suspended particles like sand. This difference makes suspended sediment concentration more representative of the water body as a whole.

As mentioned above, sediments are produced by rock weathering and make up the minerals which form rocks (Chaudhry, 2008). Due to differing compositions of various rock types, for example, igneous, sedimentary, and metamorphic, the exact nature of these sediments is dependent on where they are found and the geology of their source (Fondriest Environmental, 2014). An analysis of river water between different continents would result in different suspended bedload and solute composition. This is due to differences in lithology as well as climatic and vegetation variations (Bridge and Demicco, 2008). For example, the colder regions are normally associated with rocks and gravel while areas with high temperature and rainfall are abundant in mud and silt (Huggett, 2008). The higher the temperature of the location, the more intense is the weathering. Tropical rivers contain more soluble elements as compared to temperate and arctic rivers because high temperatures in the tropics encourage chemical weathering and lead to the leaching of soluble materials (Huggett, 2008).

Apart from the composition of sediments in river waters, there are also variations in the quantity of sediments within and between different climates. One example would be the glacial meltwater river system, where sediment concentrations have large temporal fluctuations. These variations are the result of erosional and sediment delivery processes in various catchments (Bogen and Bønsnes, 2003).

However, human activities such as poor land use practices, urbanization, and deforestation have contributed to changes in the sediment load of rivers (Hassan et al, 2017; Williams, 2012). Additionally, the main source of sediments in a catchment could be concentrated in a proportionally small area, and development in such areas could potentially lead to detrimental effects (Walling, 2005). Forestry activities such as logging, agricultural farms, and construction sites increase the suspended sediment yield when bare soil is exposed and vulnerable to erosion (Csáfordi,

2014). Loose soil is then carried via surface runoff and contributes to the sediment load in the channel.

1.3.1 SEDIMENT EROSION

Erosion is defined as the processes that remove sediments and move them to a different place. It involves sediment transport which is the study of sediment particles movement over time. It occurs in natural systems where gravity moves sediment along the surfaces which they rest on or due to fluid motions such as currents and tides (Imran, 2008). It is vital to understand sedimentation and its processes to develop more efficient flood and erosion control as well as river basin management.

Sediment load that is entrained in a flow can be classified into bedload, suspended load, and wash load. Bedload comprises sediment grains which continuously reside at the bottom of a streambed (Anderson and Anderson, 2010). The typical movement characteristics of bedload are rolling, sliding, or saltating along the streambed (Wood and Armitage, 1997). The bedload found upstream is usually different from sediments found downstream, which are rounded and reduced in size due to abrasion. On the other hand, the suspended load is the portion of sediments lifted due to turbulence in fluid flow during transportation (Anderson and Anderson, 2010). They consist of small particles ranging from clay, silt to fine sand.

Lastly, wash load comprises the finest sediment grains added at the upstream end of a channel, where they are continuously suspended without ever settling (Anderson and Anderson, 2010; Woo et al., 2011). These particles are extremely small in relation to bedload. The composition of wash load is distinctly different from bedload, consisting of clay and silts, and is often derived from other sources apart from bedrock (Biedenharn and Thorne, 2007).

1.3.2 SEDIMENT DEPOSITION

Also known as sedimentation, sediment deposition is considered the final stage of sediment transport and deposition. Deposition occurs when the sediments are added to a landform with the loss of kinetic energy, thus building up layers (Huggett, 2008). As fluid turbulence reduces, large sediments in the bedload would be deposited first, followed by the suspended load in decreasing particle sizes. Smaller sized sediments may also undergo cohesive deposition. This results in flocs forming within the water column, rather than settling at the bottom of the bed (Fondriest Environmental, 2014). Deposition can also occur in chemical environments, where the solute is being precipitated and result in landforms such as stalagmites and stalactites (Huggett, 2008).

1.4 EROSIONAL AND DEPOSITIONAL ENVIRONMENTS

Landscapes include all visible features of an area while landforms are shapes and characteristics of the land surface that results from geomorphological processes and hence landforms can be classified as subsets of landscapes (Renwick, 1992). Geomorphology is the study of landforms and the processes that shape them (Summerfield, 2001). Due to the presence of geomorphological processes such as the natural mechanisms of weathering, erosion, and deposition, surface materials and landforms on Earth's surface are constantly modified to shape dynamic landscapes that change over time.

Erosion can be simply defined as a process whereby natural forces (e.g., water, wind, gravity, or ice) wear away rocks and soil whereas deposition is a process where eroded or transported materials are being deposited due to a loss in energy. A wide variety of landforms is formed in our world today due to erosion and deposition processes. This section is focused on four types of erosional and depositional environments (Fig. 1.7), namely: (1) Fluvial—environments that are dominated by flowing water like rivers (2) Coastal—environments at the interface between land and sea, (3) Glacial—environments where ice is a major transport process, and (4) Aeolian—environments where the wind has the ability to shape Earth's surface like deserts. Erosional and depositional processes specific to these environments are discussed and the landforms that arise due to these processes are outlined in the subsequent sections.

1.4.1 FLUVIAL EROSIONAL ENVIRONMENT

Erosion in the river channel is defined as the removal and movement of material such as rocks downstream by the river and can be classified as vertical and lateral erosion. Vertical erosion causes the river channel to deepen and mainly occurs at the upper course, whereas lateral erosion widens the channel and predominantly occurs at the lower course (Stiros et al, 1999).

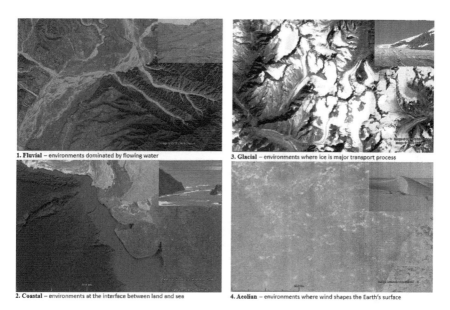

1. Fluvial – environments dominated by flowing water

3. Glacial – environments where ice is major transport process

2. Coastal – environments at the interface between land and sea

4. Aeolian – environments where wind shapes the Earth's surface

FIGURE 1.7 **(See color insert.)** Erosional and depositional environments.

There are four main types of erosional processes and they occur simultaneously throughout the whole course of the river system:

- Hydraulic action: High impact force of running water may be strong enough to loosen and dislodge rocks and small particles from the river bed and banks to erode the river channel.
- Abrasion: Eroded rock fragments transported along the river grinds against the surface of the river channel which further deepens and widens the river channel.
- Solution: Rain droplets react with carbon dioxide in the atmosphere to form a weak carbonic acid that dissolves minerals in rock on the river bed and banks.
- Attrition: Rock fragments collide with one another as they are transported along the river, and break down into smaller fragments.

The river system also plays a role in shaping the many landforms found on the Earth's surface. Considering a river system of upper, middle, and lower course, the river experiences different processes, namely erosion, transportation, and deposition, as it moves along the river course and end at the river mouth (Leopold, 2006). Some of the landforms are described in the following sections.

1.4.1.1   WATERFALLS

Waterfalls are steep and vertical flows of fast-flowing water that fall from a height and are usually located at the upper course of the river system where the gradient of the slope is steep. Waterfalls are commonly formed when the river flows through and erode a region of banded rocks with different resistance. The less resistant rocks are preferentially eroded as compared to the more resistant rocks, which are eroded at a slower pace. Over time, the eroded rock will form a small notch, promoting vertical downcutting of the river which cuts the less resistant rock to create a vertical drop. Repeated pounding of the base of the softer rock caused by the waterfall results in the formation of a plunge pool.

1.4.1.2   VALLEYS

Valleys are the low areas in between mountains. Valleys formed by vertical erosion of rivers at the upper course are usually V-shaped with steep valley sides, while further down the river system, valleys become wider and broader due to the predominance of lateral erosion. There are two main types of valleys, which are gorges and canyons. Gorges are deep, narrow, and steep-sided valleys that are formed by an active river downcutting of the valley or waterfalls receding upstream. Canyons are the extended form of gorges, with deeper and longer valleys.

1.4.1.3   MEANDERS

Meanders are the product of erosion, transportation, and deposition processes, and are usually located in the middle and lower course of the river. As the meander curves, erosion occurs at the outer river bank where the river velocity is high and there is more energy for erosion. Over time, the outer bank gets undercut by the river to form a concave slope and steep-sided bank called river cliff.

1.4.1.4 OXBOW LAKES

Oxbow lakes are the extension of meanders and are formed when the outer banks of two consecutive meanders meet to form a loop. The continuous erosion and deposition river process along the outer and inner banks, respectively, results in the meander loop being cut off from the main river and is closed off by deposited sediments to form the oxbow lake.

The combination of the three river processes results in the formation of various beautiful landforms along the river course. These landforms are not static and will change with time as the river continues to flow through (Sundborg and Rapp, 1986).

1.4.2 FLUVIAL DEPOSITIONAL ENVIRONMENT

Deposition refers to the process of eroded materials being dropped. This process occurs when the river gradient is low or when the gradient is reduced hence the river loses its energy due to the reduction in velocity thereby leading to sediments being deposited because it can no longer carry the weight of the sediments (Huggett, 2011). Deposition can occur both in the middle and lower course of the river but it mostly occurs downstream where low-lying land is found, whereby conditions are favorable for deposition (Huggett, 2011). Examples of landforms that are formed as a result of deposition include meanders, natural levees, and deltas.

As the river travels downstream, it starts to slow down in velocity. As a result, lag deposits consisting of denser and heavier particles such as boulders and gravel found within channel deeps are deposited. Transitory channel deposits consisting of coarse bedload materials such as sand, gravel, and boulders are usually deposited between low and intermediate channel flow (Sharma, 2010).

1.4.2.1 MEANDERS AND OXBOW LAKES

Meanders and oxbow lakes are formed via erosion and deposition. Channel margin deposits that are a result of lateral accretion, can be found. These smaller, fine-grained sediments such as clay, silt, and sand are deposited on the inner banks leading to the formation of point bars as the channels shift overtime during the process of meander and oxbow lake formation

(Sharma, 2010). Other types of deposits such as channel-fill deposits can also be deposited. They range from relatively coarser bedload to finely-grained deposits of oxbow lakes, from abandoned courses of meandering streams (Sharma, 2010).

1.4.2.2   NATURAL LEVEES AND BACK SWAMPS

During the formation of landforms like natural levees and backswamps, overbank floodplain deposits are deposited as a result of vertical accretion. Such deposits consist of fine-grained sediments found in the suspended load such as clay, silt, and sand. The size of natural levees would be determined by the frequency of the overbank flow and the concentration of the suspended load found in the overflow (Sharma, 2010). During floods, streams may break the levees and results in crevasse splays which are the local accumulation of relatively coarser materials deposited on the floodplains (Huggett, 2011).

1.4.2.3   DELTAS

Deltas are usually formed when sediments are deposited at the mouth of the river. This is when the sediment load is too heavy for the river to carry along with it and is deposited at the fringes. As the river loses its energy and thus velocity, the first type of sediments to be deposited first is the larger and heavier sediments and this will form the topset bed. The next type of sediments to be deposited is sediments that are of the medium size such as coarse-grained sand, forming the forest bed. The last type of sediments to be deposited is the finely-grained and lighter sediments such as clay and silt which will be carried the furthest by the river thus forming the bottomset bed (Ace Geography, 2010). From this, it is evident that the different types of deposits are involved when it comes to the formation of different depositional landforms.

1.4.3   COASTAL EROSIONAL ENVIRONMENT

The landforms in different coastal areas are formed due to a process is known as coastal erosion. There are four main ways coastal erosion can occur:

- Abrasion: The constant force of water against the shore wears it away. Particles of rock and sand in the water waves chips rocks off existing rocks in the shoreline.
- Attrition: Water causes rocks and pebbles to collide and break up.
- Solution: Acids in the sea slowly dissolves certain types of rock.
- Air compression: When air gets trapped in cracks in rocks, the pressure increases and causes cracks to expand.

A combination of the different processes in different intensities causes different landforms to be observed over time in the coastal area.

1.4.3.1 CLIFFS

The impact of waves scours away the base of the shores through the abrasion, as well as through the hydraulic action and the solution formation. Over time, a wave-cut expands and face of the cliff will be exposed with no support below.

The cliff face collapses due to its own weight, moving the cliff inwards. The processes of attrition combined with the processes of transportation then than take away the debris from the cliff. This is a repetitive process when continued over a longer period of time will create a larger wave-cut platform.

This may cause the development of the beach providing some sort of protection to the cliff (Hampton and Griggs, 2004).

1.4.3.2 HEADLANDS AND BAYS

Headlands and bays are formed due to differential erosion due to the layers of soft rocks and hard rocks (Schofield, 1967). Bays are generally defined as the sheltered zones, with low energy which is produced in the zones where there is normally weak geology, for example, they form in clay areas, where the cliff erodes at a faster rate. Headlands are the sides of the bay, generally, they are outcrops of more resistant rock like limestone which extends into the water normally perpendicular to the bay. The largely sized headlands are also known as the cape.

1.4.3.3   CAVE, ARCHES, STACKS, AND STUMPS

Caves and arches form from the lines of weaknesses in the headlands that expand due to air compression. Over time, the top portion of arches would collapse due to weight and form a stack. After further erosion by high tide or wind, a stack shortens into a stump.

1.4.4   COASTAL DEPOSITIONAL ENVIRONMENT

Coastal deposition occurs largely because of longshore drift, which is a process of sedimentary transportation. Longshore drifts occur when the prevailing winds cause waves to approach the shoreline at a constant angle, usually at an angle of 45°, carrying sand and sediments up to the beach (BBC, 2017). The backwash, however, carries the materials back down at a 90° angle which is the steepest angle possible. As the waves continue in this motion, it eventually loses energy and deposits its materials, such as sand, sediments, and shingles (Internet Geography, 2016), onto the shoreline.

Materials are moved via transportation process in four main ways, namely:

- *Traction*—large materials are dragged across the floor.
- *Saltation*—materials are bounced across the floor.
- *Suspension*—material is suspended and carried by the waves.
- *Solution*—material is dissolved and carried by the water.

Due to the processes of coastal deposition, three main landforms can be formed, namely the beach, the spit, and the tombolo (Internet Geography, 2016).

1.4.4.1   BEACH

The beach is defined as the area between the lowest and highest point reachable by the waves in any conditions and exists usually in one of the two forms, the sandy beach, and the shingles beach. Sandy beaches tend to have a gentler sloping profile as compared to the shingles beaches which have a steeper sloping profile.

The materials found along the beach can vary in size with the smallest found near the waterline and the largest found toward the back of the beach. This is attributed to a coastal erosional process known as attrition mentioned earlier, which breaks the sediments near the waterline into smaller pieces. There are also a few landforms that are formed on beaches due to coastal deposition, such as dunes and berms.

1.4.4.2   SAND DUNES

Sand dunes are small ridges of sand found at the top of the beach, usually past the maximum point reachable by waves. The sand is deposited by winds onto obstructions such as rocks, bushes, or driftwood. Over time, the dunes increase in size and may form rows perpendicular to the prevailing transverse dunes (Araújo et al., 2013; Earth Eclipse, 2017).

1.4.4.3   BERMS

Berms are essentially the terraces of the beach that are formed usually in the backshore, above the water level at high tide (Britannica, 1998).

1.4.4.4   SPIT

The waves lose energy when it encounters an obstruction and thus deposits its materials. Over a period, hooks can form from spits if materials continue to be deposited along the coastline. When this occurs, salt marshes and mudflats may also form as materials continue to be deposited in the sheltered area where waves are unable to reach.

1.4.4.5   TOMBOLO

The tombolo is essentially a spit or a bar that connects two land bodies, usually an island and the mainland. The island attached to the mainland is called the tied island. A tombolo is also known as the sandy isthmus. The term isthmus refers to a narrow piece of land joining the other larger land areas.

1.4.5   *GLACIAL EROSIONAL ENVIRONMENT*

A glacier can be simply defined as a large body of ice moving slowly under influence of gravity. Glaciers can be differentiated into two types, depending on whether they are confined (e.g., mountain alpine glaciers confined with topographic depressions) or unconfined for example continental glaciers in Antarctica (Huggett, 2011). In the interest of glacial erosional environments and landforms, the focus of this section would be on erosional landforms caused by confined mountain glaciers. The three main types of glacial erosion process are fracturing and plucking which refers to the crushing and erosion of loosened material by glaciers, abrasion which refers to the erosion of the substrate by rock debris embedded in glaciers and lastly erosion by subglacial meltwater whereby meltwater under glaciers chemically and physically weathers the material underneath it (Ritter et al., 2006). A total of nine glacial erosional landforms will be briefly introduced in this section.

1.4.5.1   *CIRQUE OR CORRIE*

A cirque or corrie is a bowl-shaped or amphitheatre-like depression produced by the glacier due to its circular motion or rotational slip. Cirques can be found in high mountain areas where there is the existence of glaciers. In small hollows, snow is collected and compacted to form ice over time. As the ice starts to move, plucking comes in action and the rocks from the back wall erode. At the base of the basin, rotational scouring (abrasion) by the ice wears down the underlying rocks. Eventually, a rock lip is formed when ice moves out of the basin. Tran is a mountain lake that is formed when rain or river water fills up the cirque.

1.4.5.2   *PATERNOSTER LAKES*

These are the series of glacial lakes generally connected by a single stream, usually formed in low depressions of a U-shaped valley. It is created by scouring a valley bed which contains rocks which are different in resistance. The Thornton lakes in North Cascades National Park, USA is an example of three connected lakes southeast of Mount Triumph forming paternoster lakes.

1.4.5.3 ARÊTE

Arête is a residual narrow knife-edged ridge that forms when two cirques erode back to back, generally gets sharpened by the freeze-thaw weathering. It is a narrow ridge produced by glacial erosion that separates the valleys. Arête is a French word meaning "edge" the German term for the same feature is "grat."

1.4.5.4 HORN

Three or more cirques may erode back to back, eventually forming a residual pyramidal peak called a horn. The Matterhorn in Zermatt, Switzerland is a classic example of a horn that is formed by cirque erosion when there is the divergence of multiple glaciers from a central point. If the horn is not covered with ice/snow and is exposed to an ice field, they are called nunatak or glacial islands.

1.4.5.5 COL

It is a glacially produced saddle (lowest point on a mountain ridge between two peaks) when cirque headwalls breach the arête that separates them. It is also known as a notch or just as a saddle. Generally speaking, a col may provide a way or pass in the mountain ranges.

1.4.5.6 HANGING VALLEY

It is a former river tributary where the floor of which meets the main glacial trough at a higher level. Initially, the glacier fills the river valley, tributary glaciers join the main glacier at heights. However, once the glacier erodes, the tributary valley will be left hanging, thereby forming a hanging valley.

1.4.5.7 GLACIAL TROUGH

It is a glaciated U-shaped valley formed when a glacier carves the valley by the action of scouring. As a result, steep and straight sides are formed together with a wide flat floor. The Geiranger Fjord in Norway is a good example of a trough carved out by glaciers. Seawater covered the floor of the U-shaped valley after glaciers receded which led to the creation of the fjord. Water still flows down from some hanging valleys into the fjord currently forming notable waterfalls such as the Seven Sisters Falls.

1.4.5.8   TRUNCATED SPURS

Throughout the formation process of a glacial trough, sides of the valley are smoothened and interlocking spurs are eroded by the moving glacier. The "worn down" spurs that are left behind are known as truncated spurs.

1.4.6   GLACIAL DEPOSITIONAL ENVIRONMENT

As glacier descends from a valley, unsorted sediments (or till) eroded from valley walls and floor are transported on the surface, within or below the glacial ice. At lower-altitude area, ablation (i.e., ice removal from the glacier by melting or evaporation) is dominant which means there is a net loss of ice.

Therefore, sediments are concentrated near the glacial toe by melt-out, which give rise to a wide range of depositional features (Evenson and Schlieder, 2017). Glacial depositional landforms can be classified based on their relative position to the glacier (supraglacial, subglacial, and marginal) and ice-flow direction (parallel, transverse, and nonoriented) (Huggett, 2011).

1.4.6.1   MORAINES

On the glacier surface, moraines are accumulated supraglacial sediments primarily derived from rock debris fallen from ice-eroded cliffs or have been dragged along by the moving glacier [Hambrey, 1994; National Snow and Ice Data Center (NSIDC), 2017]. They vary in form and position depending on the orientation of the ice flow and are less commonly found on continental ice sheets than on alpine glaciers (Hambrey, 1994; Huggett, 2011). One of the most prominent supraglacial landforms is lateral moraine which formed along the glacial margins—parallel to the ice flow (Fig. 1.8a) (Huggett, 2011). Though it is a supraglacial land-form, part of it is made of subglacial sediments that originated from the mountainsides that rubbed against the glacier. When two lateral moraines between two alpine glaciers converge, medial moraines are formed which are mostly linear but may become folded for a piedmont glacier (Hambrey, 1994). Hummocky moraines are also distinctive supraglacial features appearing as a random aggregation of hummocks made of diamicts (Huggett, 2011). Based on observation-based models, sedimentation of supraglacial debris from the stagnant glacier and dead-ice melt-out have been suggested to contribute to their formation (Braaten, 2011).

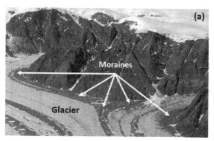 

FIGURE 1.8   Moraines (a) and drumlins (b).

1.4.6.2   DRUMLINS

Beneath the glacier involves a complex interaction between erosion and deposition of unconsolidated sediments. A prominent subglacial feature associated with continental glaciation is drumlin (Fig. 1.8b), a streamlined elongate hill with a steep stoss end and a gentle lee slope. Growing evidence suggests that its formation may have resulted partly from the deformation of till below the glacier beside streamlined movement of the ice sheets (Hambrey, 1994; Evenson and Schlieder, 2017). It is usually found in clusters known as drumlin fields, where the largest is in central-western New York, occupying an area of 225 km by 56 km which holds about 10,000 drumlins deposited by Wisconsinan Laurentide ice sheet (Hambrey, 1994).

1.4.6.3   ESKERS

Apart from ice, vast amounts of meltwater also flow on the surface, within and at the base of a glacier that creates glaciofluvial landforms made of generally well-sorted sediments. For instance, eskers are long, sinuous ridge of stratified sand and gravel, formed from sedimentation in channels below the glacier by subglacial meltwater (Huggett, 2011). They are thought to be diachronous, unlike crescentic gorges which may form in minutes or drumlins which form cumulatively over a longer time scale (Fredin et al., 2013).

Glacial studies on depositional environments have unveiled the paleoclimatic patterns and paleogeographical implications of these regions (Hambrey, 1994). In other words, glacial deposits can provide valuable data on geological past events which may help to predict future climate change. Meanwhile, glacial deposits continue to serve as important reservoirs for groundwater and sources for construction materials (Earle, 2014).

1.4.7   AEOLIAN EROSIONAL ENVIRONMENT

Wind is primarily the geomorphic agent in arid zones with very low annual precipitation and sparse vegetation such as deserts. In deserts, the annual precipitation can reach less than 300 mm which bear extensive areas of the Rocky Mountains, alluvial fans, and plateau. Aeolian erosional landforms are produced mainly by wind erosion which involves two processes— deflation and abrasion. Deflation refers to the removal of loose sand particles by wind while abrasion is the wearing effect of a rock surface by friction due to the impact of wind-carried sand particles. There are several landforms associated with wind erosion, few of the most common are: yardangs, eugenics, rock pedestals, ventifacts, desert pavements, and blowouts aka deflation hollows.

1.4.8   AEOLIAN DEPOSITIONAL ENVIRONMENT

Similar to other environments, where there is erosion, there is a deposition. In deserts, sand accumulates in a range of shapes and sizes which may occur as sheets of sands, sand dunes, or loess. For instance, sand dunes can be categorized based on their orientation such as transverse or form, for example, star, crescentic (Huggett, 2011). A few common depositional landforms are: pediments, bolson, bajada, alluvial fans, and wadi and so on.

1.5 CONCLUSION

Surface processes shape the topography of our earth, significant of which is weathering erosion and depositional processes. These processes run at varying pace at different parts of the earth and are also aided by climatic drivers. Many of these deposits and landforms which occurred in the past have been buried by later sediments. Geological investigations unravel their existence and past climatic conditions can be inferred from them. Correlation of such findings at regional scales has established the proto continents and their drift in geological time through various ice ages. Understanding of the different paleoclimates of a region has several implications for society. Reconstruction of past climate supplemented by paleontological evidence has reasoned for evolution and extinction of life forms on Earth. The spatial topology of these past landforms and their absolute dating have established the configuration of continents and oceans through the Earth's past and has gained increased significance in predicting future climate and trends.

KEYWORDS

- weathering
- erosion
- deposition
- disintegration
- wedging
- insolation
- aeolian

REFERENCES

Ace Geography. *Landforms of Fluvial Erosion and Deposition.* [Online] 2010. Available at: http://www.acegeography.com/landforms-of-fluvial-erosion-and-deposition.html (accessed Oct 27, 2017).

Allen, P. *Liberation and Flux of Sediment*; Blackwell Publishing Ltd. Arizona State University: Cambridge, MA. Weathering and Soil. *Ambio* **1997,** *15* (4), 215.

Anderson, R.; Anderson, S. *Geomorphology: The Mechanics and Chemistry of Landscapes*; Cambridge University Press: Cambridge, 2010.

Araújo, A. D., et al. Numerical Modeling of the Wind Flow Over a Transverse Dune. *Sci. Reports*, **2013**, *3* (1), 2858. DOI: 10.1038/srep02858.

Barker, W.; Welch, S.; Banfield, J. Biogeochemical Weathering of Silicate Minerals. In *Geomicrobiology: Interactions between Microbes and Minerals*; Banfield, J., Nealson, K., Eds.; Mineralogical Society of America: Washington D.C., 1997; pp 391–428.

Becker, B. Lichens. *Australian Antarctic Division,* 2016. Available at: http://www.antarctica.gov.au/about-antarctica/wildlife/plants/lichen (accessed Nov 23, 2017).

BBC, *Coastal Landscapes,* 2017. https://www.bbc.co.uk/education/guides/zsdmv9q/revision/4.

Biedenharn, D. S.; Thorne, C. R. In *Wash Load/Bed Material Load Concept in Regional Sediment Management,* Proceedings of the Eighth Federal Interagency Sedimentation Conference (8th FISC), Reno, NV, USA, 2007.

Blott, S.; Pye, K. Particle Size Scales and Classification of Sediment Types Based on Particle Size Distributions: Review and Recommended Procedures. *Sedimentology* **2012,** *59* (7), 2071–2096.

Bogen, J.; Bønsnes, T. *Erosion and Sediment Transport in High Arctic Rivers*; Svalbard, 2003.

Braaten, D. A. *Encyclopedia of Snow, Ice and Glaciers,* 2011. DOI: 10.1007/978-90-481-2642-2.

Bridge, J.; Demicco, R. Biogenic and Chemogenic Sediment Production. In *Earth Surface Processes* (Ed.), *Landforms and Sediment Deposits*; Cambridge University Press: Cambridge, 2008; pp 45–118. DOI: 10.1017/CBO9780511805516.

Britannica, *Berm* [online] 1998. Available at: https://www.britannica.com/science/berm (accessed Oct 20, 2017).

Chaudhry, M. *Open-channel Flow*, 2nded.; Springer Science+Business Media: New York, 2008. https://link-springer-com.libproxy1.nus.edu.sg/content/pdf/10.1007%2F978-0-387-68648-6.pdf (accessed Nov 1, 2017).

Chen, J.; Blume, H.; Beyer, L. Weathering of Rocks Induced by Lichen Colonization—A Review. *Catena* **2000,** *39* (2), 121–146.

Csáfordi, P. Factors Influencing Sediment Transport on the Headwater Catchments of Rák Brook, Sopron, Ph.D. Dissertation, University of West Hungary, 2014.

Danin, A.; Garty, J. Distribution of Cyanobacteria and Lichens on Hillslopes of the Negev Highlands and Their Impact on Biogenic Weathering. *Zeitschrift fur Geomorphologie* **1983,** *27*, 413–421.

Dixon, J. C. Weathering. In *Encyclopedia of Geomorphology*; Routledge: London, 2004; pp 1108–1112

Earle, S. *Physical Geology* [e-book]; BCCampus Open Textbook Project: British Columbia, 2014. Available through: https://opentextbc.ca/geology/chapter/16-4-glacial-deposition/ (accessed Nov 23, 2017).

Earth Eclipse. *What is Sand Dune?* [online] 2017. Available at https://www.eartheclipse.com/geology/sand-dune-formation-types.html (accessed Oct 22, 2017).

Evans, I. Salt Crystallisation and Rock Weathering: A Review. *Revue de Geomorphologie Dynamique* **1970,** *19*, 153–177.

Evenson, E. B.; Schlieder, G. *Glacial Landform* [online]; Encyclopaedia Britannica, Inc, 2017. Available at: https://www.britannica.com/science/glacial-landform/Depositional-landforms#toc49776 (accessed Nov 23, 2017).

Fondriest Environmental. Sediment Transport and Deposition. *Fundamentals of Environmental Measurements,* 2014. http://www.fondriest.com/environmental-measurements/parameters/hydrology/sediment-transport-deposition/ (accessed Nov 1, 2017).

Fredin, O., et al. Glacial Landforms and Quaternary Landscape Development in Norway, *Quaternary Geology of Norway Geological Survey of Norway Special Publication,* 2013, *13* (2013), pp 5–25.

Fry, E. The Mechanical Action of Crustaceous Lichens on Substrata of Shale, Schist, Gneiss, Limestone and Obsidian, *Ann. Bot.* **1927,** *41,* 437–469.

Goldich, S. A Study of Rock Weathering, *J. Geol.* **1938,** *46,* 17–58.

Goudie, A. Insolation Weathering. In *Encyclopedia of Geomorphology;* Routledge: London, 2004; pp 566–567.

Griggs, D. The Factor of Fatigue in Rock Exfoliation. *J. Geol.* **1936,** *44* (7), 783–796.

Gupta, A. *Weathering in the Tropics;* Cambridge University Press: Cambridge, 2011.

Hall, K. The Role of Thermal Stress Fatigue in the Breakdown of Rock in Cold Regions. *Geomorphology* **1999,** *31,* 47–63.

Hallet, B.; Walder, J.; Stubbs, C. Weathering by Segregation Ice Growth in Microcracks at Sustained Subzero Temperatures: Verification from an Experiemental Study Using Acoustic Emissions. *Permafrost Periglac. Processes* **1991,** *2,* 283–300.

Hambrey, M. *Glacial Environments* [e-book]; CRC Press LLC: Florida, 1994. Available at: Google Books http://booksgoogle.com (accessed Nov 23, 2017).

Hampton, M.; Griggs, G. *Formation, Evolution, and Stability of Coastal Cliffs—Status and Trends* [online] 2004. Available at: https://pubs.usgs.gov/pp/pp1693/ (accessed Nov 2, 2017).

Hassan, M.; Roberge, L.; Church, M.; More, M.; Donner, S. Leach, J.; Ali, K. What are the Contemporary Sources of Sediment in the Mississippi River? *Geophys. Res. Lett.* **2017,** *44* (17), 8919–8924.

Huang, W.; Keller, W. Organic Acids as Agents of Chemical Weathering of Silicate Minerals. *Nature (Phys. Sci.)* **1972,** *239,* 149–151.

Huggett, R. The Geomorphic System. *Fundamentals of Geomorphology,* 2nd ed.; Routledge: London, England, 2008; pp 31–48.

Huggett, R. *Fundamentals of Geomorphology;* Routledge: Abingdon, 2011.

Huggett, R. J. *Fundamentals of Geomorphology* [e-book]; Routledge: New York, 2011. Available through: Taylor & Francis Group Library Website http://www.tandfebooks. com.libproxy1.nus.edu.sg/doi/view/10.4324/9780203860083 (accessed Nov 23, 2017).

Imran, J. Sediment Transport. In *Open-Channel Flow;* Chaudhry M. H., Ed.; Springer: Boston, MA, 2008; pp 453–477.

Internet Geography. *Coastal Transportation* [online], 2016. Available at: http://www.geog-raphy.learnontheinternet.co.uk/topics/whatiscoastaltransportation.html (accessed on Oct 20, 2017).

Schofield, J. C. Sand Movement at Mangatawhiri Spit and Little Omaha Bay. *New Zeal. J. Geol. Geophys.,* **1967,** *10* (3), 697–721. DOI: 10.1080/00288306.1967.10431087.

Leopold, L. B. *A View of the River;* Harvard University Press, 2006.

Merrill, G. *A Treatise on Rocks, Rock Weathering and Soils;* Macmillan: New York, 1897.

Moses, C.; Smith, B. A Note on the Role of Collema Auriforma in Solution Basin, 1993.

National Snow and Ice Date Center [NSIDC]. [online] 2017. Available at: https://nsidc.org/cryosphere/glaciers/questions/climate.html (accessed Nov 24, 2017).

Nicholson, D. Granular Disintegration. In *Encyclopedia of Geomorphology;* Routledge: London, 2004, 493–494.

Oakland Geology Salt Weathering, 2014. https://oaklandgeology.wordpress.com/2014/05/11/salt-weathering-tafoni/ (accessed Oct 31, 2017).

Peltier, L. C. The Geographic Cycle in Periglacial Regions as it is Related to Climatic Geomorphology. *Ann. Assoc. Am. Geogr.* **1950,** *40* (3), 214–236.

Polo, E. Frost Weathering, 2014. Available at: http://edupolovalcarcel.blogspot.sg/2014/11/weathering.html (accessed Nov 30, 2017).

Puente, M.; Bashan, Y.; Li, C.; Lebsky, V. Microbial Populations and Activities in the Rhizoplane of Rock-Weathering Desert Plants. I. Root Colonization and Weathering of Igneous Rocks. *Plant Biol.* **2004,** *6* (5), 629–642.

Renwick, W. Equilibrium, Disequilibrium, and Nonequilibrium Landforms in the Landscape. *Geomorphology* **1992,** *5* (3–5), 265–276.

Ritter, D. F.; Kochel, R. C.; Miller, J. R. *Process Geomorphology. IIE Trans.* **2006,** *2.* http://doi.org/10.1080/07408170108936849.

Robinson, D.; Williams, R. B. Accelerated Weathering of a Sandstone in the High Atlas Mountains of Morocco by an Epilithic Lichen. *Zeitschrift fur Geomorphologie* **2000,** *44,* 513–528.

Sharma, V. K. *Introduction to Process Geomorphology* [e book]; CRC Press: Boca Raton, 2010. Available at: Google Books (accessed Nov 30, 2017).

Sperling, C. H. B.; Cooke, R. Laboratory Simulation of Rock Weathering by Salt Crystallization and Hydration Processes in Hot, Arid Environments. *Earth Surf. Process. Land.* **1985,** *10* (6), 541–555.

Sthos, S. C.; Darkas, N., Moutsoulas, M. River Erosion and Landscape Reconstruction in Epirus: Methodology and Results. *Brit. School Athens Stud.* **1999,** *3,* 108–114.

Strakhov, N. *Principles of Lithogenesis*; Oliver and Boyd: Edinburgh, 1967.

Summerfield, M. *Global Geomorphology*; Longman Scientific & Technical: Harlow, Essex, 1991.

Summerfield, M. *Geomorphology and Global Tectonics*; Wiley & Sons: Chichester [u.a.], 2001.

Sundborg, Å.; Rapp, A. Erosion and Sedimentation by Water: Problems and Prospects. In *Hydration. Encyclopedia of Geomorphology*; Trudgull, S., Ed.; Routledge: London, 1986; pp 534–535.

Tsytovich, N. *The Mechanics of Frozen Ground*; Scripta Book: Washington D.C., 1975.

Viles, H. A. Organic Weathering. In *Encyclopedia of Geomorphology*, 2004, 728.

Walling, D. Tracing Suspended Sediment Sources in Catchments and River Systems. *Sci. Total Environ.* **2005,** *344* (1–3), 159–184.

Williams, M. River Sediments. *Philos. Trans. R. Soc. A Math. Phys. Eng. Sci.* **2012,** *370* (1966), 2093–2122.

Wilson, M. J.; Jones, D. Lichen Weathering of Minerals: Implications for Pedogenesis. *Geol. Soc. Lond., Spec. Publ.* **1983,** *11* (1), 5–12. http://sp.lyellcollection.org/content/11/1/5. abstract.

Winkler, E.; Wilhelm, E. Saltburst by Hydration Pressures in Architectural Stone in Urban Atmosphere. *Geol. Soc. Am. Bull.* **1970,** *81,* 567–572.

Woo, H. S.; Pierre, Y. J.; Everett, V. R. Washload and Fine Sediment Load. *J. Hydraul. Eng.* **1986,** *112* (6), 541–45.

Wood, P. J.; Armitage, P. D. Biological Effects of Fine Sediment in the Lotic Environment. *Environ. Manage.* **1997,** *21* (2), 203–217.

CHAPTER 2

Minerals and Rock-Forming Processes

MUHAMMAD NAWAZ[1,*], FARHA SATTAR[2], and
SANDEEP NARAYAN KUNDU[3]

[1]*Department of Geography, National University of Singapore, Singapore*

[2]*College of Education, Charles Darwin University, Darwin, Australia*

[3]*Department of Civil & Environmental Engineering, National University of Singapore, Singapore*

Corresponding author. E-mail: geomn@nus.edu.sg

ABSTRACT

Minerals and rocks are formed in a variety of ways, and in several conditions. Rocks are usually an aggregate of minerals and non-minerals. This chapter examines how mineral and rocks are formed and describe the different types of minerals and rock-forming processes. Mineral-forming processes cover minerals from molten materials, minerals from solutions, metamorphism and minerals formation, and weathering and mineral formation. Rock-forming processes describe the igneous processes and rocks, sedimentary processes and rocks, and metamorphic process and rocks.

2.1 INTRODUCTION

The surface of the earth is made up of rocks which contain different kinds of minerals. Many of these minerals are rarely seen because they are buried deep within the earth. A mineral is a naturally occurring, inorganic solid compound, which has a fixed crystalline structure and specific chemical composition. Minerals are collections of one or more elements neatly stacked together; there exist thousands of possible combinations of these elements, which results in a myriad of minerals. Igneous, sedimentary, and metamorphic are mainly made up of minerals. This chapter covers the mineral-forming processes and also describes the processes forming the igneous, sedimentary, and metamorphic rocks.

2.2   MINERAL-FORMING PROCESSES

Minerals can form in a variety of condition, such as cooling of magma, precipitation from aqueous solutions, and metamorphic and weathering processes. Through identifying the minerals present in a rock, a geologist will able to connect the circumstances resulting in the mineral and hence understand the conditions necessary for the formation of these rocks. This information can be used as a source for other areas with similar conditions that could be potential/predicted sources of the minerals of interest.

Crystallization is the key process in the mineral formation. Minerals form either from crystallization of molten material such as magma or lava or from the crystallization of solutions (Fig. 2.1).

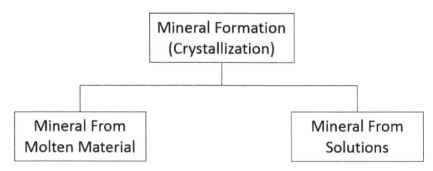

FIGURE 2.1   Mineral formation through crystallization.

This section covers the main mineral-forming processes that give birth to the majority of minerals on Earth.

2.2.1   MINERALS FROM MOLTEN MATERIALS

The convection within Earth's mantle causes magma to rise from the upper mantle toward the crust and the rising magma slowly starts to cool. Convection transfers hot material upward and cooler material downward.

When the hot material reaches new condition and new temperature below the melting point, the hot liquids crystallize. This is the most common way by which minerals form on the Earth. For this process, the temperature is important factor that determine the type of the minerals formed. This factor is based on Bowen's series (Klein, 1999), which shows the type of minerals formed at specific temperatures during the magma-cooling process (Fig. 2.2).

At higher temperatures, minerals are formed from mafic magma, of low viscosity and low silica content, and intermediate magma. There are two starting branches of progression for mineral formation here.

First, the discontinuous branch involves minerals that contain iron and magnesium silicates in their chemical composition. At high temperatures of around 1200–1300°C, olivine is formed. When allowed to cool further, olivine would react with the magma and recrystallize to form pyroxene. Again, given time and abundance of silica, the continuous cooling of magma would allow pyroxene to recrystallize into amphibole and eventually biotite. This is due to the mineral's attempt to achieve stability by changing its internal crystalline lattice, as its temperature decreases.

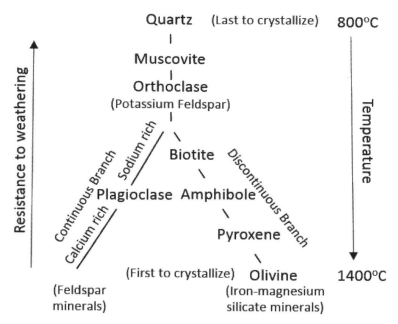

FIGURE 2.2 Bowen's reaction series.

Source: Modified from Klein (1999).

The continuous branch involves plagioclase feldspars that evolve from calcium-rich to sodium-rich. Compared to the discontinuous branch, the basic crystalline lattice of the mineral does not change. As the temperature decreases, calcium plagioclase of a high melting point will partially react with magma to form plagioclase containing a higher sodium composition, until eventually, and sodium plagioclase of lower melting point is formed. Due to this, plagioclase crystals usually have a calcium-rich core with a more sodium-rich exterior.

Next, at lower temperatures around 750–800°C, these two branches merge to follow the same progression. Also, the minerals are formed from intermediate and felsic magma, which are of high viscosity. These minerals contain potassium and are high in silica content. This composition gives them a lighter color, compared to minerals from mafic magma. They are also more stable, due to their lower melting points, and thus less susceptible to weathering.

Time is important as it affects the size of the mineral crystal formed, by determining how long the atoms have to bond and form a more orderly arrangement. However, this does not affect the chemical composition of the minerals themselves.

When the molten magma cools slowly, such as when it is still beneath the Earth's crust, the atoms will have more time to bond. This allows the mineral crystals to slowly grow (Morse and Wang, 1996) before it rises above the surface and completely solidifies. Conversely, when the cooling process is rapid, such as when the magma is on the Earth's surface, the crystals will not have time to grow and will remain relatively small. Additionally, when this process is too quick, the crystals will not have time to form at all.

As the magma oozes out onto the Earth's crust from below, due to various tectonic activities, it is called lava. The lava continues to cool, at an even faster rate due to the lower temperatures at the top of the lithosphere, and more minerals will crystallize.

2.2.2 MINERALS FROM SOLUTIONS

Aqueous solutions can precipitate to form minerals both on and underneath the surface of Earth. At the surface, when water in an aqueous solution evaporates due to surrounding pressure and temperature, it leaves behind minerals. Beneath the surface, the groundwater reacts with rocks to form dissolved minerals. Flowing through cracks in the rocks, the groundwater cools and deposits solid minerals which will form vein (Desonie, 2012).

One of the minerals formed from aqueous precipitation is halite. A chloride salt, halite is usually formed on the beds of ancient seas and salt lakes. As the water of these water bodies slowly evaporated, solid deposits of halite are left behind (Gardiner, 2003). One example is The Great Salt Lake in Utah, United States, where water has evaporated with minerals and salt left behind, forming the lake.

pH is an important factor that could affect the type of minerals precipitated from different water bodies on Earth. At different pH, different chemical species may be present in the solution, which can alter the chemical composition of minerals formed from that solution. For instance, leaching of lead is very common at extremely high and low pH. This causes a water body to have a higher concentration of dissolved lead ions, which leaves behind lead-containing minerals after evaporation (Sauvé et al., 1998). Also, at high pH, the alkalinity in the water body may be due to the interactions of the water body with carbonate rocks (Fondriest Environmental Inc., 2013). As such, the precipitation of alkaline water could produce more carbonate-based minerals instead.

Gases play an important role in the precipitation of minerals from solutions. When magma rises to the surface, gases trapped in magma are released due to decreasing pressure. Other sources of gas could also be due to the heating of groundwater. As the mineral-containing gas ascends, it could leach more minerals out of the surrounding cooling magma/rocks (Zhu et al., 2011). When it meets water in the aquifers, it will dissolve and this increases the concentration of minerals in the body of waters. As such, when the mineral cools, this would promote the precipitation of minerals from aqueous solutions.

Also, magmatic gas can escape into the atmosphere through various forms such as fumaroles, that is, openings in the Earth's crust and volcanic vents. When gases containing minerals flow through fractures and fissures in the volcanic structures, it cools and precipitates (Patterson, 2003) to form minerals along these openings, which creates mineral veins. This is dominant in volcanic regions, as various volcanic gases are emitted. An example of this is the precipitation of sulfur, which is emitted from fumaroles and forms sulfur-rich minerals.

2.2.3 *METAMORPHISM AND MINERALS FORMATION*

Metamorphism is the formation of new minerals from existing minerals through recrystallization under high temperatures and pressure. Recrystallization is the rearrangement of atoms in an existing mineral to a novel one (Rebrovic, 2013). However, its chemical composition may or may not remain unchanged, as additional elements or compounds may be included from the surroundings. The atoms and ions in the minerals migrate due to high temperatures and may form new arrangements due to high pressure. These novel minerals formed via metamorphism in metamorphic rocks are also known as index minerals which are used to determine the extent of metamorphism in a rock (Alden, 2017).

There are various types of metamorphism (Nelson, 2017). First, thermal/contact metamorphism occurs when intrusive igneous rocks get heated under low pressure and produce nonfoliated metamorphic rocks so-called hornfels. It happens in the contact aureoles surrounding an intrusion of magma and higher grade zones occur in the inner parts of aureole. The intrusion can be passive, forceful, and fluid which will produce a different kind of minerals and rocks.

Second, dynamic/cataclastic metamorphism is due to mechanical deformation by sudden large shear forces, such as earthquakes, and it is restricted to shear zone at low temperature. This process only takes a few seconds and is common at plate boundaries.

Third, regional metamorphism affects rocks in an extensive area attributing to mountain building and produces foliated rocks caused by strong pressure from colliding plates. It is the most common type of metamorphism and it is a process that spans millions of years. While the area is undergoing regional metamorphism, other metamorphic processes such as contact or dynamic can concurrently be in operation.

Besides the three main metamorphism mentioned above, there is also hydrothermal metamorphism, when rocks are modified at high temperature and pressure by hydrothermal fluids; burial metamorphism, when sedimentary rocks are buried deep down in Earth and are subjected to pressure and temperature changes; and shock/impact metamorphism, when an extraterrestrial collision with the Earth imparts high amounts of force on minerals.

One of the minerals formed through metamorphisms is the garnet. Commonly due to regional metamorphism, the chemical bonds within minerals like mica are broken due to the heat and pressure causing the minerals to recrystallize giving more stability in the new conditions. Simultaneously, additional elements can be included, which results in variations of the garnet formed. For example, metamorphism of clay sediments results in Almandine, an iron-rich garnet form. Conversely, metamorphism of carbonate rocks results in calcium-rich garnets like Andradite.

Temperature and pressure set the conditions for the different types of metamorphism and the resulting minerals formed. According to the geothermal gradient, the temperature increases while going deeper in the Earth. In addition, temperatures are higher near magma intrusions. Higher temperatures could also speed up the rate of metamorphism, as it speeds up the motion and interactions of chemical molecules. Additionally, pressure increases with depth too. The greater the pressure exerted, the greater the extent of metamorphism exhibited, if the metamorphism is pressure-dependent such as dynamic metamorphism.

2.2.4 *WEATHERING AND MINERALS FORMATION*

On the Earth's surface, minerals are more exposed and vulnerable to other natural phenomena, such as erosion and wind. This is because minerals that are formed intrusively might not be stable as it is uplifted above the crust, due to the different conditions above and beneath the Earth's crust (Nelson, 2017). On the surface, the minerals experience lower temperatures and pressures, and they are more exposed to agents of weathering. Two main agents of weathering are water and oxygen (O_2). As such, the presence of water and oxygen are factors that affect the weathering process. To adapt to these new conditions, the unstable minerals undergo various chemical weathering processes whereby they are broken down into smaller fragments are reformed with different chemical compositions, due to the addition of other chemical species. There are many ways a mineral can be chemically weathered, such as hydrolysis, oxidation, carbonation, and hydration.

Hydrolysis: Due to the abundance of water, the H^+ and OH^- ions present in the water will react with ions in the minerals. This breaks down the minerals to produce new substances. Feldspar, a silicate, is very susceptible to hydrolysis and ionizes to form more stable clay minerals such as kaolinite.

Oxidation: The free oxygen in the atmosphere reacts with the minerals to form oxides, or hydroxides when water is also present. Iron is particularly susceptible to this process, whereby siderite reacts with oxygen to form hematite and undergoes hydration to form limonite.

Carbonation: When rainwater reacts with CO_2 in the air, it produces a weak carbonic acid which can alter carbonate minerals such as calcite, a constituent of limestone. The reaction produces soluble substances (Olajire, 2013) which are then removed as solutions. They can then reprecipitate as formations such as stalactites.

Hydration: Minerals are altered by incorporating water (Snellings et al., 2011) into their molecular structure. It also causes physical stresses which can result in disintegration. This process prepares minerals for further alteration by oxidation and carbonation, as it allows more surface area and greater ease of any ion transfer.

The stability of the minerals on the Earth's surface is largely dependent on the temperature at which it is formed (Nelson, 2017). Minerals formed at higher temperatures will be more unstable at the surface while minerals formed at lower temperatures are more stable and will undergo lesser chemical weathering (Fig. 2.2).

2.3   ROCK-FORMING PROCESSES

The igneous, sedimentary, and metamorphic are the three main types of rocks found on Earth. The rock cycle is the mechanism that models the evolution and recycling of these rocks throughout geological time and is a vital part in explaining the composition of the Earth. Through these several natural igneous, sedimentary, and metamorphic processes, we can see how rocks can be changed from one form to another.

2.4   IGNEOUS PROCESSES AND ROCKS

Out of the three types of rocks found on Earth, igneous rocks are the most abundant (Evers, 2015). Formation of igneous rocks is closely related to tectonic activities which involve igneous processes like cooling and solidification of magma underneath the crust or lava on the surface. The varieties of igneous rocks and their textures depend on these igneous processes which involve the original magma composition, the rate of cooling, location of cooling, and the surrounding host rocks. Texture and composition are the two most important properties form distinguishing and classifying igneous rocks. Igneous rocks may have the same chemical composition but when formed under different processes, they will have different texture and appearance. Rocks formed under intrusive processes are called plutonic rocks. Whereas, rocks formed under extrusive processes are called volcanic rocks. Some examples of rocks with the same chemical compositions but having different physical appearance due to different solidification conditions are: granite (plutonic) and rhyolite (volcanic) which have a felsic composition; gabbro (plutonic) and basalt (volcanic) which have a mafic composition. There are two types of processes that form igneous rocks, namely, intrusive and extrusive processes.

2.4.1 INTRUSIVE PROCESSES

Intrusive igneous rocks are formed by magma solidifying under the Earth's crust. Temperature remains relatively high throughout the cooling process as the crust acts as a blanket to insulate the magma underneath. Thus, the solidification process is slower. As magma is very concentrated in some minerals, the slow rate of solidification allows more time for the aggregation of these minerals which results in an appearance of large crystals within the solidified rock. As a result, coarse grains and large crystals visible to the naked eye are usually observed on intrusive igneous rocks. In other words, intrusive igneous rocks are known to have a phaneritic texture. Common intrusive igneous rocks include granite, diorite, and gabbro (Martin, 2017).

During an eruption, the lava that surfaces above the Earth's crust is just a small fraction of the total magma that has reached the level of the crust. Most of the total mass of magma is usually left to solidify underneath the Earth's crust. The intrusive rocks are then eventually exposed after long periods of erosion on the Earth's surface. Intrusive igneous rocks may also form igneous bodies such as batholiths, dykes, and sills (Fig. 2.3).

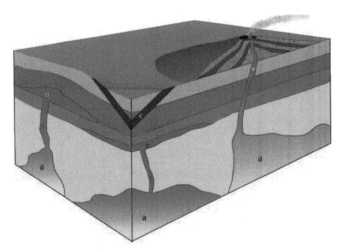

FIGURE 2.3 **(See color insert.)** Different types of the plutons: (a) stocks, (b) sill, (c) dyke, (d) laccolith, (e) pipe, and (f) pipes or dykes.

Source: Reprinted from Earle (2014). https://opentextbc.ca/geology/chapter/3-5-intrusive-igneous-bodies/ https://creativecommons.org/licenses/by/4.0/

Batholiths are very large bodies of igneous rocks that cooled intrusively. They can have a diameter of up to a few hundred kilometers and are usually irregularly shaped. Due to the intrusion process of batholiths, high temperature and pressure cause metamorphic processes to be observed in the rocks surrounding the batholiths.

Dykes and sills are the most common igneous bodies exposed on the earth surfaces. They are generally long and thin mostly similar in size and usually identified by the difference in color compared to surrounding rocks. They are also younger than surrounding rocks. However, the differences lie in their geometry. Dykes are squeezed vertically between country rocks. Hence, they are vertical intrusions. Dykes are often found together and they generally have basaltic composition. Mackenzie Dyke Swarm in Northwest Territories of Canada is the world's largest dyke swarm.

In contrast, sills are horizontal intrusions that are a parallel between layers of country rocks. They are harder to identify as they are often confused with lava flow. The main difference between them would be the bubble holes that appear in lava flow due to volcanic gases escaping out from it. The Great Whin Sill and the Drumadoon on the Isle of Arran are where sills can be found.

The intrusive igneous rock bodies introduced above can be categorized based on how they intrude the surrounding rocks. Concordant igneous rock bodies are aligned with the preexisting rock layers whereas discordant igneous rock bodies cut across preexisting layers of rock. Sills are concordant igneous rock bodies while dykes and batholiths are discordant igneous rock bodies.

2.4.2 EXTRUSIVE PROCESSES

On the other hand, extrusive igneous rocks are formed from lava that solidified above the Earth's crust. As lava has a much higher temperature than the air and water in the Earth's atmosphere, upon exposure to air or water, lava will cool very rapidly. The rapid cooling process does not allow enough time for minerals in the lava to aggregate before complete solidification of the rock. Thus, extrusive igneous rocks have very fine grains and small crystals which may only be visible under a microscope. As a result, extrusive igneous rocks are known to have aphanitic texture. Common extrusive igneous rocks include rhyolite, andesite, and basalt.

Although igneous rocks are commonly classified according to their composition and texture, and they are often present in plutonic and volcanic pairs, there are some other types of extrusive igneous rocks that do not fall into these categories such as *tuff, obsidian, scoria,* and *pumice* (King, 2017).

Tuff is a type of extrusive igneous rock that is formed from the compaction and consolidation of materials (mostly volcanic ash) released from volcanic eruptions. Although the formation of tuff does not involve magma cooling and solidification, it is closely related to igneous activities and therefore considered as a type of igneous rocks.

When lava cools very quickly such that no crystals could form, *obsidian* is formed. It has a glassy appearance and therefore it is often called as "volcanic glass." This extrusive igneous rock can be found at various sites around the world. However, it is mainly found in areas with recent volcanic activity. This is because weathering processes and heat destroy old obsidian rocks overtime.

Scoria is formed when magma that contains a high concentration of dissolved gases is ejected onto the surface of the Earth. Due to the decrease in pressure as magma rises, dissolved gases escape as bubbles in the magma. If these gas bubbles do not manage to escape before the lava solidifies, they would be trapped in the rock and the resultant rock would appear to have cavities.

Pumice is an igneous rock similar to scoria. The difference between them is the magma that forms them. Pumice is formed from rhyolite magma, while scoria is from basaltic magma. Also, pumice usually has more holes than scoria such that the structure of the rocks between the bubbles is very thin. This is also the reason why pumice rocks float on water while scoria sinks.

Porphyritic rocks are extrusive igneous rocks with both fine and large crystals. Thus, porphyritic rocks have physical characteristics of both intrusive and extrusive igneous rocks. Crystallization occurs in the magma before a volcanic eruption. This results in large crystals present in the magma before the eruption. When the eruption occurs, the large crystals are ejected onto the Earth's surface along with lava which cools rapidly and forms extrusive igneous rocks that contain the large crystals.

2.4.3 COMPOSITION-BASED CLASSIFICATION

Igneous rocks classified according to their composition, include four categories, namely, ultramafic, mafic, intermediate, and felsic. The composition of igneous rocks is closely related to the composition of magma from which the rocks are formed, and the composition of magma varies greatly at different locations. The geographical location, for example, subduction zones, mid-oceanic ridges, and hot spots, where the igneous rocks originate largely determine its mineral composition, and therefore its color.

At divergent plate boundaries, igneous rocks are formed. The divergence of the plate boundaries causes the faults and fractures in the crust, hence allowing magma from the upper mantle to leak out. The upper mantle mostly consists of mafic magma containing ferromagnesian silicates which means it has high iron, calcium, and magnesium and low silica content. The lava then quickly solidifies on the seafloor to form basalt. The rest of the magma that is trapped under the oceanic surface slowly cools to form gabbro. Basalt and gabbro are mainly made up of plagioclase and pyroxene minerals. Basalt generally ranges from dark grey to black in color. It has 45–50% SiO_2, and is mostly aphanitic but also can be porphyritic. Gabbro is generally greenish, dark-colored, and is phaneritic or pegmatitic. It has similar silica content as basalt as they are from the same magma source. The basalt formed at the mid oceanic-ridge is called mid-oceanic-ridge basalt (MORB) due to the distinct compositional range of the basalt formed. Due to the spreading of the plates, the youngest MORB is the nearest to the ridge, whereas the oldest MORB is the furthest from the ridge (Jahns and Kudo, 2017). This forms the ocean floor hence, most of the oceanic crust is made of MORB. For example, the mid-Atlantic ridge which is formed by the divergence of the North American and Eurasian plate is made of MORB.

Submarine eruptions from submarine volcanoes due to hotspots also form mafic igneous rocks. These hotspots are formed due to mantle plumes which are the upwelling of the hot material from the lower mantle. These hotspots cause igneous activities which are not due to plate boundaries and movement. One prominent example is Hawaii, where submarine volcanoes gradually developed into islands due to hotspot volcanism, which does not lie near any plate boundary. These submarine volcanoes or volcanic islands are generally made up of ocean island basalt (OIB) (Deschamps et al., 2011). These OIBs are more enriched in elements than MORBs. This is due to the source of the magma for the mantle plume is the lower mantle reservoir, and the magma from the lower mantle is more enriched than that of the upper mantle. Basalt and gabbro are formed at hot spots as well.

2.4.4   OCCURRENCES OF IGNEOUS ROCKS

Igneous rocks are also found in subduction zones where an oceanic plate is subducted beneath another oceanic plate or continental plate. As the subducted plate descends toward the asthenosphere, the temperature increases and eventually reach a point that the subducted oceanic plate melts into magma. Furthermore, as plenty of water is present in subduction zones, it can provide an extra amount of thermal energy to break the chemical bonds in minerals, resulting in a lower melting point for the crustal rocks. Therefore, extensive igneous activities are happening at subduction zones, a large amount of magma is formed and solidified, either beneath or above the ground surface, to form igneous rocks.

Different from its counterparts formed at mid-oceanic ridges which contain less silica, magma, and igneous rocks formed at subduction zones are mostly intermediate (53–65% silica) and felsic (>65%) in composition. The reason why mafic oceanic crust rocks produce magma of higher silica content is partial melting. According to Bowen's reaction series, the minerals that melt first are silica-rich, including quartz, potassium feldspar, and sodium-rich plagioclase, whereas pyroxene, olivine, and calcium-rich plagioclase, which have less silica content, melt at a later stage. Hence, when the mafic oceanic crust is only partially melted, the magma should have a higher silica content than the source rock, thus having an intermediate or felsic composition. In addition, some of the silica-rich sediments and sedimentary rocks rested on the ocean bed are carried down as well by the subducted oceanic plate, which can melt under the high temperature and contribute to the silica content of produced magma. The composition of the magma could further be changed by a process called assimilation. It occurs when the rising magma comes into contact with rocks in the continental crust, and due to the extremely high temperature of the magma (up to 1300°C), the surrounding continental rocks, which have a higher silica content than oceanic crust, could melt completely or partially to further contribute silica-rich materials.

Igneous rocks of intermediate composition (53–65% silica), formed at subduction zones, are andesite and diorite. They have equivalent compositions, but andesite is extrusive (fine-grained) and diorite is its intrusive (coarse-grained) counterpart. The predominant mineral composing these two rocks is plagioclase feldspars (~60% by volume), and typically the rest ferromagnesian components are amphibole or biotite. The name "andesite" is derived due to its extensive presence in the Andes Mountains in South America. The Andes Mountains is a continental volcanic arc formed by the subduction; the Nazca Plate is being drawn down the South American Plate. Other than Andes Mountains, andesite is also commonly found in the Cascade Range in western North America, occurring at a convergent boundary.

Felsic magma (>65% silica) solidifies and form rhyolite and granite. Rhyolite is extrusive and fine-grained and granite is plutonic and coarse-grained. They are largely composed of nonferromagnesian minerals like quartz, potassium feldspars, and sodium-rich plagioclase. A minute amount of biotite and amphibole could also be present. Granite is the most commonly found intrusive igneous rock so far in the world. Most of them were formed at convergent plate boundaries in the past beneath the ground, and when the mountains are uplifted and eroded over time, the plutonic granitic rocks are exposed and appear on the surface (Wicander and Monroe, 2006).

2.5 SEDIMENTARY PROCESSES AND ROCKS

Sedimentary rock is the product of lithification of sediments. Generally, sediments are attained from the mechanical or chemical weathering of other rocks. Eventually, these sediments get transported and deposited before undergoing lithification to form rock. Among all classes of rocks, sedimentary rock constitutes the most common rock type on Earth which covers approximately 73% of Earth's surface (Wilkinson et al., 2009). This section describes the formation of three classes of sedimentary rock: clastic, chemical, and organic. This subdivision is based on their formation processes and constituents. Each formation process will be discussed with examples of rock type to illustrate the detailed mechanism involved in the formation of sedimentary rocks.

2.5.1 CLASTIC SEDIMENTARY ROCKS

Clasts are rock fragments or grains which are formed from weathering of the preexisting rocks. These clasts are transported and deposited in suitable sedimentary environments where they are compacted and cemented to form clastic sedimentary rocks. Examples of clastic sedimentary rocks are mudrocks (i.e., shale, slate, mudstone, claystone, and siltstone), sandstone, conglomerate, and breccia (Fig. 2.4).

(a) Mudrock with bivalve impressions, Cretaceous Nanaimo Group, Browns River, Vancouver Island

(b) Coarse sandstone with cross-bedding, Cambrian Tapeats Formation Chino Valley, Arizona

(c) Conglomerate with imbricate (aligned, tilted down to the left) cobbles, Cretaceous Geoffrey Formation, Hornby Island, BC

(d) Sedimentary breccia, the Pre-Cambrian Toby Formation, east of Castlegar, BC (image is approx. 1 m across)

FIGURE 2.4   Mudrock, sandstone, conglomerate, and breccias.

Source: Reprinted from Earle (2014). https://opentextbc.ca/geology/chapter/6-1-clastic-sedimentary-rocks/ https://creativecommons.org/licenses/by/4.0/

Mechanical weathering is the first step of clastic rock-forming processes. Mechanical weathering is a physical means of the breakdown of rock which is aided by frost wedging, exfoliation cracks, or gravity-aided rockfall. Larger clasts rare then transported by natural agents like wind, ice, or water. During transportation, these clasts erode further through attrition and abrasion processes into smaller clasts and into different shapes and sizes. When transportation ceases they are eventually deposited at a location where over time these sediments compact forming horizontal layers. The lower layers are overburdened with high pressure from the layers above, thus become compacted. Water from the interstitial spaces between sediments is eventually excluded as the spaces diminish due to compaction. Subsequently, cementation occurs when dissolved chemicals remained in the interstitial water become saturated and precipitate, gelling the sediments together to form rocks. Calcite, silica, and hematite are amongst the common cementing agents. This process of compaction and cementation is also known as lithification (Nelson, 2015).

The geometric properties of the sediments provide information on its agent of transportation and deposition. The shape of the sediments can also provide a hint on the shape of the original rock, while flatness of the sediment provides information on lithology. Generally, thin-bedded rocks yield flat pebbles, while larger rocks produce more equidimensional fragments.

There are various sedimentary environments that exist on Earth, each associated with unique sediment types. Generally, sedimentary environments can be divided into two: terrestrial and marine. Lakes and rivers are examples of terrestrial sedimentary environments, whereas deltas and coastal beaches are examples of marine sedimentary environments (Nelson, 2015). By identifying the components of sedimentary rock, the environmental conditions of the rock origin can be deduced.

2.5.1.1   CONGLOMERATES AND BRECCIA

Conglomerates and breccias are made up of large-sized clasts called gravel which have dimensions greater than 2 mm. Conglomerates are coarse-grained rock with rounded gravel. They are generally classified into two types, extraformational and intraformational. Like conglomerates, breccias are coarse clastic sedimentary rocks made up of gravel as well. However, their difference lies in the fact that breccias are angular, while conglomerates are rounded. The angular shape of breccia indicates that the clasts are not transported for a large distance, and are deposited near to their origin.

2.5.1.2   SANDSTONES

Sandstones are made up of several components, such as: major minerals, accessory minerals, rock fragments, and chemical cement. The major minerals come in the form of sand particles with sizes ranging between 1/16 and 2 mm, which make up the framework fractions (King, 2017).

The dominant major mineral in the sandstone is quartz, making up 65% of the average sandstone (Boggs, 1995). Other major minerals can also be present, such as feldspars and clay minerals. However, feldspars are generally less dominant as they are susceptible to chemical and mechanical weathering due to their chemically reactive properties and soft texture. The accessory minerals are the minerals having the average abundance of less than 1–2% in sedimentary rocks but can go as high as 2–3% in sandstones (Boggs, 1995). Mica is an example of a common accessory mineral.

Rock fragments are also present in sandstone and these are pieces of existing rock that have yet disintegrated. A sandstone can have rock fragment proportion ranging from 0% to 95%, but average sandstone should have rock fragment of around 15–20% (Boggs, 1995). Rock fragments are important in studies of sedimentary rocks as they are reliable indicators of the source. Chemical cement can be present at varying amounts in the sandstone as part of the matrix. These can be silicate minerals, carbonate minerals, iron oxide minerals, and sulfate minerals.

Sandstones can be classified as quartz arenites, feldspathic arenites, or lithic arenites. If the matrix can be recognized, the terms quartz wacke, feldspathic wacke, and lithic wacke are used (Williams et al., 1982).

2.5.1.3   SHALE

Shale is mainly composed of clay and silt particles and form at places that are abundant with fine sediment and with low water energy, which allows settling of fine silt and clay.

2.5.2   CHEMICAL SEDIMENTARY ROCKS

Unlike the clastic sedimentary rocks, the chemical sedimentary rocks are not composed of individual grains of sediments. Instead, these rocks are made up of one or more material of specific chemical composition compounds in solid form which is accumulated through chemical processes from a solution. The processes of deposition involve primarily precipitation and evaporation (Tucker, 2009).

Evaporation is the main process in the formation of chemical sedimentary rocks (Mackenzie, 1971). It can be most commonly observed in large inland salt lakes, central regions of remote oceanic basins as well as bordering on sea margins (Greensmith, 2012), where common deposits are that of halite and dolomite. The chemical compounds or minerals that make up chemical sedimentary rocks must be first dissolved in water (Berner, 1971). Water evaporates from these solutions containing the dissolved sediments, thereby concentrating the sediments to its saturation point. As water continues to evaporate, the sediments, now called evaporites start to precipitate and crystallize to form rock (Berner, 1971).

Conditions favoring the accumulation and maintenance of salt deposits include high evaporation rates in less humid environments, high surface temperatures above 15°C, coupled with little rainfall below 200 mm per year, and slow replenishment of water (Greensmith, 2012). However, if the efflux of bottom brine exceeds or is equal to the amount of salt entering these water bodies, the total net salt concentration will either decrease or remain the same. Hence, the saturation point is unable to be reached and common evaporite minerals will be unable to precipitated (Berner, 1971).

Once conditions are favorable for the accumulation of salt deposits, large-scale salt deposition is likely to occur which can lead to the precipitation of various salts, with some common and others uncommon. Common ones usually comprise the sulfates and chlorides of sodium, magnesium, calcium, and potassium (Sloss, 1953). For example, simple standard salts such as halite (sodium chloride) and anhydrite (calcium sulfate), as well as double salts like polyhalite and other complicated amalgamations are typically formed (Sloss, 1953).

Precipitation of evaporites happens in two instances, one within the denser and lowermost layers of water bodies encompassing concentrated salt solutions called brine, and another at and immediately beneath interface of water and sediment (Sonnenfeld, 1984). The precipitation of salt occurs in sequence with the increment of salt concentrations.

Due to the specific minerals in chemical sedimentary rocks, it can be further categorized into different groups according to its composition. There are five major rock types in chemical sedimentary rocks, namely, carbonates, evaporites, cherts, ironstones, and phosphorites (Boggs, 2009). In this discussion, we will concentrate on carbonate rocks, the most abundant chemical sedimentary rock (Boggs, 2009).

2.5.2.1 CARBONATE ROCKS

According to Boggs (2009), "Carbonate rocks make up one-fifth to one-quarter of sedimentary rocks in the stratigraphic records." Carbonate rocks can be further classified into two major types: limestone and dolomites. Limestone is a commonly occurring chemical sedimentary rock. Limestone is a class of rocks that has a higher percentage of carbonate, with at least 80% of carbonate constituents (Pettijohn, 1975). The carbonate constituents of limestone are primarily composed of calcite or aragonite, both of which are naturally occurring crystal forms of calcium carbonate ($CaCO_3$). Although some limestones are formed through biochemical precipitation, it can also be formed through abiotic means. A few examples are: travertine, tufa, and caliche. Travertine is formed through precipitation of carbonate minerals, usually around the hot springs which are heated geothermally, whereas tufa is developed through the process of precipitation in ambient water. Caliche, on the other hand, is formed via evaporation of lime-rich water when it is drawn to the surface of the soil in semiarid regions (Pettijohn, 1975).

The dolomites are a variety of limestone where calcium magnesium carbonate is more than 50% of the carbonate content of the rock (Pettijohn, 1975). The dolomites can be formed in three ways: dolomitization, dolomite cementation, and precipitation (Boggs, 2009). Dolomitization is the process where limestone is converted to dolomite through the replacement of calcium carbonate with calcium magnesium carbonate (Boggs, 2009). Dolomite can also precipitate at low temperatures where there is an abundance of organic material (Roberts et al., 2013).

Limestone forms the majority of the mass of limestone caves such as the Jenolan Caves in New South Wales, Australia. Out of the many structures in the cave, one of them is a dripstone. They form when water seeps through the limestone and becomes concentrated with calcium carbonate, leading to its deposition with the evaporation of water and release of carbon dioxide (A Dictionary of Geology and Earth Sciences, 2013). This thin layer of calcium carbonate formed hangs from the roof of the cave and is gradually accreted as more and more calcium carbonate precipitates out from the water. Eventually, long icicle-like bodies hanging from the roof called stalactites takes shape, while droplets which accumulate on the floor become pillars known as stalagmites.

2.5.3   ORGANIC SEDIMENTS AND ROCKS

Organic sedimentary rock is formed by organic materials originated from the living organism and they are generally classified according to their composition as well as formation mechanism. Some of these rocks such as coral limestone may sometimes be classified under chemical sedimentary rocks for it being chemically precipitated; however, as the synthesizing process involved biological host, in this chapter, it will be classified as an organic sedimentary rock.

Generally, there are two main mechanisms by which an organic sedimentary rock can form, namely, cementation or compaction of fossils which give rise to bioclastic rocks; and crystallization of precipitates of biologic origins which give rise to crystalline organic rocks. Bioclastic rocks can consist of either calcite or carbon as their main composition, forming limestones and coals that are generally of animal origin for the formal and plant origin for the latter, whereas crystalline organic rocks are almost exclusively composed of calcite, as they are normally secreted as exoskeleton by aquatic creatures. Three examples, namely, coals, coral reefs, and coquina will be examined to understand each type of formation mechanism.

2.5.3.1   COAL

Coal is a carbon bioclastic rock formed from the build-up of remnants of plant debris that is eventually processed for energy fuel known as petroleum. There are two main types of coal, namely, humic coals and sapropelic coals. Humic coals come from the remains of dense woody vegetation (Tucker, 2009) found in moist aerobic swamps and bogs (O'Keefe et al., 2013). Under this environment, the vegetation undergoes petrification to produce peat (O'Keefe, 2013). Sapropelic coals, on the other hand, come from spores, algae (Speight, 2012), and nonwoody plant material in anaerobic environments found in still water. Under this environment, these plant materials undergo putrefaction to give rise to organic mud (O'Keefe, 2013).

In addition to the source of coal, the organic constituents, also known as maceral composition, of both humic and sapropelic coals could be used to differentiate both coals. While the marceral composition of humic coals is made of mainly vitrinite or inertinite, sapropelic coals are mainly made of liptinite; with sporinite, in cannel coal or with alginite, in boghead coal (O'Keefe, 2013).

2.5.3.2 BIOCLASTIC CARBONATES

Coquina is classified as a detrital limestone, as it is formed by loosely cemented shell fragments of molluscs that are generally larger than 2 mm. Aragonite contributes to the main composition in coquina as is amongst the most common compound animals used to construct their shells. These shell fragments are normally transported and deposited in the high-energy environment with shallow water where waves break, for example, beaches and raised banks, in which during the process the shells were broken into tiny pieces. The shell fragments accumulated on the beach over thousands of years. Over time, cementation will occur at the lower layers when acidic rainwater dissolves calcium in the shell to form calcium carbonate, which will then be carried to the bottom, and eventually precipitate and solidify in between the shell fragments, thereby gelling them together to form coquina.

Unlike most limestones, coquina has soft and porous texture due to the lack of cement matrix. In the past, the soft textured coquina turned out to be an ideal material for fort building, as it allows absorption of cannon balls instead of shatter and fracture. One very famous example would be the walls of the Castillo de San Marcos in Florida, USA.

Other than calcium carbonate, silicate minerals, especially quartz can sometimes be found to form a part of coquina. However, carbonate grains should be of the majority of the rock component or otherwise it should be classified as carbonate sandstone or conglomerate.

2.5.3.3 CRYSTALLINE ORGANIC ROCK

Rather than being considered as an ever-enriching marine ecosystem, coral reefs, in the geological sense, is classified as a type of organic sedimentary rock. It consists of mainly aragonite, one of the crystal forms of calcium carbonate that are produced by homotypic corals, or more commonly known as stony corals.

Coral reefs are built by layer upon layer of polyp calcareous skeleton resulting in a calcite crystalline organic rock. Stony corals secrete calcium carbonate as a cuplike skeleton periodically. When the corals die, their skeleton remains and serve as the foundation to which new polyp larvae can attach and grow, and the cycle of reef-building begin again.

Typically, coral reefs are found in shallow water of tropical region with the average temperature between 21°C and 29°C. The strict geological distribution is due to the coral's symbiotic relationship with Zooxanthellae, a microscopic alga that lives in the gastroderm, the gastrovascular lining of polyps which gives the corals its typical reddish-brown or green color (Carbonniere, 2008).

Though the calcification mechanism in corals is not yet fully understood, Zooxanthellae are known to be essential for the coral calcification process in a few ways. First, this microscopic alga conducts photosynthesis to generate energy. As much as 98% of the energy produced will be passed to the polyps as an essential energy source for the saturate calcium and carbonate ions through active transport (Constantz, 1986; Carbonniere, 2008). More importantly, through photosynthesis, Zooxanthellae actively remove CO_2, thereby shifting the carbonate equilibrium and drives the precipitation.

2.6   METAMORPHIC PROCESSES AND ROCKS

Metamorphic rocks make up almost 12% of the Earth's current land masses. These rocks are a result of the geological process of metamorphism, where the minerals within the rock react to form new minerals of varying chemical composition under the influence of high temperature and pressure (Fig. 2.5). Metamorphism is applicable to all rock, but the most significant changes take place deep within the Earth's crust but some changes can be seen closer to the surface.

Metamorphism is not a zero-sum process; it occurs gradually over a spectrum of intensity. In order for the rock to be considered as metamorphic, there must be a clear distinction in the observable features of the rock from its original form and these changes would have to occur in physical or chemical conditions not usually present on the surface of the Earth. Metamorphism can be divided into a lower limit and an upper limit.

At the lower limit of metamorphism, rocks undergo metamorphic processes that are similar to that of lithification, otherwise known as diagenesis. During diagenesis, buried sediment particles are lithified and changed into the solid sedimentary rock at temperatures and pressures lower than that of metamorphism. In order for a rock to undergo meta-morphism, the heat and pressure must be sufficient to surpass the stage of solidifying the rock to cause recrystallization to occur. This limit is not fixed either as it depends upon the main composition of the rock being metamorphosed. For example, rocks dominated by silicate materials usually witness mineral growth at 150 ± 50°C, and this mineral growth categorizes the process as metamorphic in nature.

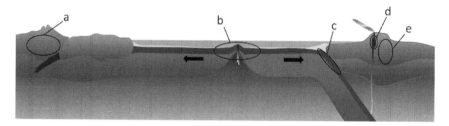

FIGURE 2.5 Metamorphism and plate tectonics: (a) regional metamorphism related to mountain building, (b) regional metamorphism of oceanic crust, (c) regional metamorphism of oceanic crustal rocks, (d) contact metamorphism, and (e) regional metamorphism related to mountain building.

Source: Reprinted from Earle (2014). https://opentextbc.ca/geology/chapter/7-3-plate-tec-tonics-and-metamorphism/ https://creativecommons.org/licenses/by/4.0/

At the upper limit of metamorphism, rocks do not just undergo defor-mation but rather minerals melt completely and reform under much higher temperature and pressure. The melting temperature ranges from 650°C to more than 1100°C, depending on the bulk mineralogical composition and amount of water present in the rock. This is most applicable to the metamorphosis of rocks such as granite that have a large composition (by volume) of igneous rock.

2.6.1 AGENTS OF METAMORPHISM

Metamorphic rocks differ depending on the environment they are formed in and the various metamorphic processes they are subjected to. There are four main agents of metamorphism that give rise to the variety of rocks, namely, temperature, pressure, fluid, and stress. These agents are suffi-cient to change the structure, mineral composition, and texture of already existing rocks. In most formation processes, rocks are subjected to more than one agent.

2.6.1.1   TEMPERATURE

Temperature plays a significant role in rock metamorphosis. Rocks are exposed to high temperatures within the Earth's crust when there are magma intrusions or when plates converge toward each other causing subduction. Intense heat causes the atoms in the rocks to vibrate vigorously and form a new bond with other atoms, resulting in a change in the structure of the rocks. Atoms can also be rearranged among grains through a process known as solid-state diffusion, resulting in a change in the mineral composition of rocks (Marshak, 2013). The degree of metamorphism is dependent on the geothermal gradient, which relates increasing temperature to depth. The deeper the rock is located, the higher the temperature is; and since, pressure also increases with depth, temperature and pressure are also directly proportionate. Temperature increases at an average rate of 20–30°C/km.

2.6.1.2   LITHOSTRATIGRAPHIC PRESSURE

As mentioned, pressure increases when the depth increases. This is known as the lithostatic pressure which is a uniform field of pressure that is caused by the weight of overlying rock. The further the rock is in the Earth's crust, the higher the pressure it experiences, at a rate of 10 kilobars at every 33 km of burial. When pressure increases, the atoms in the rock are forced closer to each other forming tighter bonds with a more stable structure. Grains are also forced together which reduces the pore space, forming denser rocks. Therefore, given that temperature and pressure are dependent on depth, rocks formed at different depths possess very different properties as the degree of metamorphism they were subjected to differ.

2.6.1.3   TECTONIC STRESS

Stress, unlike lithostatic pressure, is a nonuniform pressure exerted more intensely in one direction. Stress can be experienced in two forms, compressive and shearing. Compressive stress flattens the rocks and it is commonly experienced at convergent plate boundaries. Shearing, on the other hand, is experienced when rocks are slid parallel to one another and this is common along transform plate boundaries. The grinding action causes the rocks to fragment into finer grains (Nelson, 2014). Both forms of stress can change the shape and texture of the rocks without breaking them by realigning the minerals, resulting in rock foliation.

2.6.1.4   HYDROTHERMAL FLUID

Hydrothermal fluids are the last common agent of rock metamorphosis. Hydrothermal fluid can be found in existing rock or may seep up into the rock from surrounding magma or down into the rock from overlying groundwater. Hydrothermal fluids chemically react with rock and accelerate metamorphic reactions as atoms can travel faster in liquid mediums which gives rise to new minerals in the rocks (Marshak, 2013).

2.6.2   TYPES OF METAMORPHISM

Metamorphosis of rocks at different locations on the Earth's crust differ due to varying temperature and pressure conditions and hydrothermal fluids availability all of which aid chemical reactions and transformation of minerals. Based on the agents and their dominance in the process of metamorphism, three main types of metamorphism are identified, namely, hydrothermal metamorphism, dynamic metamorphism, and contact metamorphism.

2.6.2.1   HYDROTHERMAL METAMORPHISM

Availability of hot water aids low temperature and pressure metamorphism at shallower depths. Hydrothermal metamorphism occurs in rocks near the surface where there is an intense activity of hot water. Some example locations are the Yellowstone National Park in the Northwestern United States, the Salton Sea in California, and Wairakei in New Zealand. Circulation of groundwater in the vicinity of igneous hotspots provides the necessary heat exchange and migration of chemical elements in solution which aid metamorphism. Hydrothermal fluid circulation facilitates the transportation of chemical elements forming mineral veins in existing rock fractures and also altering the surrounding rocks. Such type of metamorphism is prevalent around porphyries and often results in enrichment of metals oxides and sulfides of ore grade.

2.6.2.2   CONTACT METAMORPHISM

Contact metamorphism occurs in geologically active regions when existing protolith is exposed to intense heat but low pressure over a relatively short period of time approximately 1000 years. As a result of convection currents in the asthenosphere, magma rises into cracks or fractures in the Earth's crust forming magma intrusions which heats up the surrounding rock.

These intrusive magma chambers are known as plutons. Since the heat from the magma is conducted to a small area surrounding the intrusion, metamorphism is restricted to the zone of rock around the intrusion, otherwise known as the area most directly in contact with the magma chamber. The belt of metamorphic rock that forms around the pluton is known as aureole and the larger the intrusion is, the larger the aureole. The grade of metamorphism increases toward the magma intrusion. High-grade rocks form immediately adjacent to the aureole where the temperatures are highest over 600°C, due to immediate contact with the magma chamber and progressively lower grade rocks form further away from the aureole. The country rock furthest away from the aureole remains unaltered. It is possible for hydrothermal fluid to act on contact metamorphic rocks as fluid may seep into the rocks from the magma intrusion, accelerating the rate of metamorphosis (Marshak, 2013).

Contact metamorphism is responsible for groups of rocks like hornfels, which are hard, compact, highly metamorphosed rock usually found in the inner aureoles. They consist mainly of oxide and silicate minerals, not necessarily in equilibrium and is likened to the igneous rock granite. Hornfels comprises rocks of various grain sizes.

2.6.2.3   DYNAMIC METAMORPHISM

In the upper part of the lithosphere, rocks are cooler due to their distance away from the heat in the mantle. Hence, over here, rocks are most commonly deformed rather than melted. This deformation process occurs through the fracturing of rock, giving rise to faults, cracks, and joint. This process is otherwise known as dynamic metamorphism. Dynamic metamorphism involves rock metamorphosis at lower temperatures as compared to the contact metamorphism but at higher pressures, mainly at fault lines and plate boundaries where mechanical stress dominates. Dynamic metamorphism occurs predominantly along transform boundaries where sudden tectonic plate activity gives rise to extremely high pressures (Nelson, 2014). A significant example of the transform plate boundary is the boundary between the Pacific plate and the North American plate that creates the San Andreas Fault in the California, USA (Mason, 1990).

A unique feature of rocks resulting from dynamic metamorphism is their textures, as it is very telling of the mechanical processes of deformation and fragmentation that they undergo. Metamorphic rocks formed by dynamic metamorphism are subjected to shearing. When two plates slide past each other along a fault, the rocks are milled into fine grains caused by the brittle grinding action, forming a fault breccia. Most dynamic metamorphic rocks with informative features are usually made up of a single mineral and not an assemblage of minerals in equilibrium. Examples of minerals commonly found in dynamic metamorphic rocks are quartz and calcite (Callegari, 2007). At temperatures above 300°C, the rock undergoes intense ductile shearing and eventually forms mylonite which has foliation that roughly parallels the fault.

2.6.2.4 REGIONAL METAMORPHISM

Metamorphism occurring at a regional scale is known as regional metamorphism. This process usually occurs at levels of the lithosphere nearer to the surface, along convergent plate boundaries where either two continental plates collide, resulting in Himalayan-type regional metamorphism or when continental and oceanic plates collide, resulting in Andean-type regional metamorphism. As the name suggests, the metamorphic processes usually occur in areas of mountain-building as it has for millions of years. When two plates collide, as both are buoyant and of almost equal density, the crust is pushed together, compressing the rocks at the boundary, and buckling upwards causing the crust to thicken and form Fold Mountains. When the plates collide, a portion of the crusts are pushed downward and are subjected to high temperature from the mantle which may cause some of the rock to melt. During convergence, the immense pressure also causes faults and cracks to form in the crust through which the molten rock heats up and rises to the surrounding rock, causing small-scale contact metamorphism. The degree of metamorphism reduces further away from the plate boundaries (Winkler, 1965).

When oceanic and continental crusts collide, the oceanic crust subducts beneath the continental crust because of its higher density. The portion of crust that subducts melts because of the heat from the mantle forming a large pool of magma known as batholith, which once cools, adds to the thickness of the continental crust. Some of the magma rises through cracks in the overlying continental crust causing contact metamorphism. Dynamic metamorphism occurs when the plates collide and slide by each other. Regional metamorphism therefore not exclusive, it is a combination of contact and dynamic metamorphism that occurs over a large area.

Since regional metamorphism occurs over such a large area, it tends to produce many rocks of a varying grade. Higher grade metamorphic rock forms closer to the zone of heating deeper in the lithosphere while lower grade metamorphic rock forms further away from the zone of heating, closer to the surface. Some examples of regional metamorphic rock in ascending order of grade include slate, phyllite, schist, gneiss, and migmatite.

2.6.3   CLASSIFICATION OF METAMORPHIC ROCKS

As metamorphic rocks result from other rock types (sedimentary or igneous) that have been subjected to intense heat and pressure, these large-scale tectono-thermal processes cause them to develop different properties from its precursor rock material. This can be attributed to the fact that when the rock melts and reforms, it undergoes processes such as recrystallization, where old minerals may be partially or entirely replaced by new ones. This confers upon it certain properties that can be classified by looking at its texture, mineral composition, and foliation (Bucher, 2011).

There is no correct way of classifying metamorphic rocks. Metamorphic rocks can be grouped in three main ways. First, they can be classified based on their grade. The most common factors to take into account when classifying the rocks this way is based on their mineral compositions as well as rock structure and how susceptible it is to transformation under temperature and pressure (Bucher, 2011).

Second, they can be grouped according to their nomenclature. Metamorphic rocks are usually named with prefix-type modifiers such that the resulting rock from the metamorphic process can be easily identified. This is also dependent on its mineralogical or chemical composition, and its structure. Third, they can be categorized based on their descriptions, by outlining a certain set of properties that constitute a particular class of rock.

Metamorphic rocks are usually classified into two main categories; foliated and nonfoliated metamorphic rocks.

2.6.3.1 FOLIATED ROCKS

Foliation occurs when the immense pressure or shear stress flattens and elongates the rock during recrystallization such that the new minerals aggregate in a flaky, sheet-like structure along a plane. Foliated rocks usually form as a result of metamorphism under high-pressure conditions such as deeper in the lithosphere, near the mantle. Due to the high temperature and pressure, new minerals form from the existing ones and they will be forced to form in the directing perpendicular to the area of applied pressure. Foliated rocks exhibit several characteristics such as cleavage, schistosity, and a gneissose structure.

Cleavage refers to the type of foliation where the rock exhibits split along subparallel planes, giving rise to a platy structure with lines of weaknesses, and the flat surface perpendicular to the direction in which the pressure is applied (Earle, 2014).

Examples of cleaved metamorphic rocks include slates, phyllites, schist, and gneiss (Fig. 2.6). Slates are fine-grained, low-grade metamorphic rocks that possess a well-developed horizontal planar element. They usually occur at a low temperature of less than 350°C and a low pressure. Common minerals that makeup slates include quartz, muscovite, and chlorite but the mineral crystals are so tiny that they usually can only be seen under high magnification. Slates are dense and brittle and usually exists in gray, black, red, or green. Freshly cleaved surfaces are usually dull. Phyllites can be likened to slates for most properties, including their well-developed cleavages. However, phyllites differ from slates in terms of the lustrous sheen of their cleaved surfaces and are also made up of slightly coarser-grain minerals such as phyllosilicates. It is common to find phyllites with both foliation and lineation (Earle, 2014).

Schistosity refers to the property where foliation is due to the arrangement of relatively large crystals of inequant mineral grains that aggregate along a horizontal plane. The grain of these minerals is usually coarse enough to be seen with the naked eye. This term may be used to describe rocks with strong lineation rather than just a schistose structure. Schist is an example of a metamorphic rock exhibiting this property. It is of mediocre metamorphic grade and usually occurs at a temperature of

350–600°C. Since schistosity is a rather broad characteristic, slates and phyllites can also be considered types of schists. Schists are rich in the minerals chlorite, muscovite, biotite, and mica.

a) Slate, near to Golden, BC   b) Phyllite, location unknown

c) Schist, location unknown   d) Gneiss from the Victoria area, BC

FIGURE 2.6   Foliated metamorphic rocks: the slate, phyllite, schist, and gneiss.

Source: Reprinted from Earle (2014). https://opentextbc.ca/geology/chapter/7-2-classification-of-metamorphic-rocks/ https://creativecommons.org/licenses/by/4.0/

Another structure found in metamorphic rocks is the *gneissose* structure. Gneissose rocks are usually coarse-grained, with poorly developed schistosity with the minerals tending to separate into bands of different colors. An example would be the metamorphic rock Gneiss. These high-grade metamorphic rocks usually occur at a high temperature of ~600°C and at high pressure as well. The common minerals that makeup gneiss are feldspar and quartz, which are lighter in color and biotite and hornblende which are darker in color. The color differences are apparent in bands that form across the rock.

2.6.3.2 NONFOLIATED METAMORPHIC ROCK

Nonfoliated rocks are usually formed in an environment without direct pressure such as near the surface of the Earth's crust. For example, granofels in which the minerals show no preferred orientation, thus it lacks any of the structures mentioned above. Hornfels are a type of compact and fine-grained granofel that usually results from contact metamorphism.

There are other forms of nonfoliated metamorphic rocks which form under low-pressure conditions examples of these rocks are marble, quartzite, greenschist/greenstone, and amphibolite. Marble is usually formed from the metamorphism of limestone or dolostone and is made up of the minerals calcite and dolostone. Quartzite is usually formed from metamorphism of the sandstone rich in quartz (quartz arenite). Greenstone and greenschist are formed under low temperature and low-pressure environments and have the same chemistry and mineralogy. Greenstone is nonfoliated whereas greenschist is schistose and foliated. Amphibolite are massive metamorphic rocks which mainly comprises the minerals hornblende (>40%) and plagioclase. Migmatites on the other hand are composite silicate rock formed out of a granitic photolith through high grade metamorphic processes (Bucher, 2011).

2.7   CONCLUSION

Minerals and rock form in a variety of ways, and can form in a variety of conditions. Rocks are usually an aggregate of minerals and nonminerals, but they are mainly composed of different minerals. Minerals form by the process of crystallization. The convection within the Earth's mantle causes magma to rise from the upper mantle toward the crust and rising magma slowly starts to cool. Convection transfers hot material upward and cooler material downward. When the hot material reaches new condition, and new temperature below the melting point, the hot liquids crystallize. This is the most common way minerals form on the Earth. As the magma oozes out onto the Earth's crust from below, the lava continues to cool, at an even faster rate due to the lower temperatures at the top of the lithosphere, and more minerals will crystallize. Aqueous solutions can precipitate to form minerals both on and underneath the surface of Earth. At the surface, when water in an aqueous solution evaporates due to surrounding pressure and temperature, it leaves behind minerals. Metamorphism is the formation of new minerals from existing minerals through recrystallization under high temperatures and pressure.

Intrusive igneous rocks are formed by magma solidifying under the Earth's crust and extrusive igneous rocks are formed by the solidified lava over the Earth's crust. Igneous rocks are commonly classified according to two properties, namely, texture, and composition, both of which are closely related to the formation processes. There are two types of processes that form igneous rocks, namely, intrusive and extrusive processes. Igneous rocks may have the same chemical composition but when formed under different processes, they will have different texture and appearance.

Sedimentary rocks are formed by the process lithification and compaction of the sediments produced from the other exiting rocks. Generally, sediments are attained from the mechanical, chemical, or biological weathering of other rocks. Eventually, these sediments get transported and deposited before undergoing lithification to form rock. Sedimentary can be formed in various ways, and not just through physical means such as the lithification of individual sediments. Other ways include chemical processes such as evaporation and precipitation of minerals or biochemical ways like secretion from living organisms.

Metamorphic rocks result from other rock types including the sedimentary, igneous, or other metamorphic rocks that have been subjected to intense heat and pressure, these large-scale tectono-thermal processes cause them to develop different properties from its precursor rock material. The four main agents of metamorphism that give rise to the variety of rocks are, namely, temperature, pressure, fluid, and stress. These agents are sufficient to change the structure, mineral composition, and texture of already existing rocks. In most formation processes, rocks are subjected to more than one agent. Contact, dynamic, and regional metamorphism are the three types of metamorphic processes that contribute to the formation of metamorphic rocks. Foliation occurs when the immense pressure or shear stress flattens and elongates the rock during recrystallization such that the new minerals aggregate in a flaky, sheet-like structure along a plane. Nonfoliated rocks are usually formed in an environment where there is no direct pressure for example such as near the crust.

The study of metamorphic rocks allows us to know about the great physical and chemical changes that take place deep within the Earth. The mineralogical composition of the rock allows geologists to evaluate under what conditions, for example, temperature and pressure, the rock had been formed.

KEYWORDS

- **crystallization**
- **metamorphism**
- **weathering**
- **clastic**
- **organic**
- **tectonic**

REFERENCES

Alden, A. *What Are Index Minerals?* [Online] 2017. https://www.thoughtco.com/what-are-index-minerals-1440840 (accessed Oct 31, 2017).

Berner, R. A. *Principles of Chemical Sedimentology*; McGraw-Hill Inc.: USA, 1971.

Boggs, J. S. *Petrology of Sedimentary Rocks*; Cambridge University Press: Cambridge, 2009. DOI: 10.1017/CBO9780511626487.

Boggs, S. *Principles of Sedimentology and Stratigraphy*, 2nd ed; Prentice Hall: Englewood Cliffs, NJ, 1995; pp 168, 178, 182.

Bucher, K. *Petrogenesis of Metamorphic Rocks*; Springer Publishing: New York, 2011.

Callegari, E. N. P. *Contact Metamorphic Rocks*; Systematics of Metamorphic Rocks, 2007.

Carbonnière, A. The French Polynesian Atolls. *Hermatypic Coral: The Role of the Zooxanthellae* [Online] 2008. http://www.atolls- polynesie.ird.fr/glossaire/ukzooxan.htm (accessed Nov 10, 2017).

Constantz, B. R. Coral Skeleton Construction: A Physiochemically Dominated Process. *PALAIOS* **1986,** *1* (2), 152–157.

Deschamps, F.; Kaminski, E.; Tackley, P. A Deep Mantle Origin for the Primitive Signature of Ocean Island Basalt. *Nat. Geosci.* **2011,** *4* (12), 879–882.

Desonie, D. *Mineral Formation* [Online] 2012. https://www.ck12.org/earth- science/Mineral-Formation/lesson/Mineral-Formation-HS-ES/ (accessed Nov 16, 2017).

Earle, S., Physical Geology; BCcampus Open Textbook Project: British Columbia [Online] 2014. https://open.bccampus.ca/2015/09/03/new-open-textbook-physical-geology/ (accessed Dec 7, 2017).

Evers, J. National Geographic Society. National Geographic Society [Online], 2015. https://www.nationalgeographic.org/encyclopedia/crust/ (accessed Nov 16, 2017).

Fondriest Environmental Inc. pH of Water: Environmental Measurement Systems [Online] 2013. http://www.fondriest.com/environmental- measurements/parameters/water-quality/ph/ (accessed Nov 10, 2017).

Gardiner, L., Halite: Windows to the Universe [Online] 2003. https://www.windows2uni-verse.org/earth/geology/min_halite.html (accessed Nov 15, 2017).

Greensmith, J. *Petrology of the Sedimentary Rocks*; Springer Science & Business Media, 2012. https://www.springer.com/gp/book/9789401196406 (accessed Nov 12, 2017).

Jahns, R.; Kudo, A. Igneous Rock: Encyclopedia Britannica [Online], 2017. https://www.britannica.com/science/igneous-rock/Intrusive-igneous-rocks#ref618885 (accessed Nov 1, 2017).

King, H. Rocks: Geology.com [Online] 2017. http://geology.com/rocks/ (accessed Nov 2, 2017).

Klein, C. *Manual of Mineralogy: (After James D. Dana)*, 21st ed. revised; Wiley: New York, 1999.

Marshak, S. *Essentials of Geology*, 4th ed.; W.W. Norton & Company, Inc: New York, 2013.

Martin, V. Differences Between Extrusive and Intrusive Rocks [Online] 2017. https://sciencing.com/differences-between-extrusive-intrusive-rocks-10017336.html (accessed Nov 2, 2017).

Mason, R. *Petrology of the Metamorphic Rocks*; Springer: Dordrecht, 1990.

Morse, J. W.; Wang, Q. Factors Influencing the Grain Size Distribution of Authigenic Minerals. *Am. J. Sci.* **1996,** *296*, 989–1003.

Nelson, M. *Metamorphic Rocks*; Gareth Stevens Publishing: New York, 2014.

Nelson, S. A. *Types of Metamorphism*; 2017. http://www.tulane.edu/~sanelson/eens212/typesmetamorph.htm (accessed Nov 16, 2017).

Nelson, S. A. Sediment and Sedimentary Rocks. Sedimentary Rocks [Online], 2015. http://www.tulane.edu/~sanelson/eens1110/sedrx.htm (accessed Nov 2, 2017).

O'keefe, J. M.; Bechtel, A.; Christanis, K.; Dai, S.; Dimichele, W. A.; Eble, C. F.; Esterle, J. S.; Mastalerz, M.; Raymond, A. L.; Valentim, B. V.; Wagner, N. J. On the Fundamental Difference Between Coal Rank and Coal Type. *Int. J. Coal Geol.* **2013,** *118, 58 87.*

Olajire, A. A. A Review of Mineral Carbonation Technology in Sequestration of CO_2, 2013.

Pettijohn, F. J. *Sedimentary Rocks*, 2nd ed.; Harper: New York, 1975.

Rebrovic, L. Under Pressure: Metamorphic Rocks, 2013. https://leakuhta.wordpress.com/2013/11/18/under-pressure-metamorphic-rocks/ (accessed Nov 11, 2017).

Roberts, J. A.; Kenward, P. A.; Fowle, D. A.; Goldstein, R. H.; González, L. A.; Moore, D. S. Surface Chemistry Allows for Abiotic Precipitation of Dolomite at Low Temperature. *Proc. Natl. Acad. Sci. U.S.A.* **2013,** *110* (36); 14540–14545. http://doi.org/10.1073/pnas.1305403110.

Sauvé, S.; McBride, M.; Hendershot, W. Lead Phosphate Solubility in Water and Soil Suspensions. *Environ. Sci. Technol.* **1998,** *32*, 388–393.

Sloss, L. L. The Significance of Evaporites. *J. Sediment. Res.* **1953,** *23* (3), 143-161

Snellings, R.; Mertens, G.; Elsen, J. Supplementary Cementitious Materials. *Cement Concrete Res.* **2011,** *41*, 1244–1256.

Sonnenfeld, P. *Brines and Evaporites*; Academic Press, 1984.

Speight, J. G. *The Chemistry and Technology of Coal*; CRC Press, 2012.

Tucker, M. E. *Sedimentary Petrology: An Introduction to the Origin of Sedimentary Rocks*; John Wiley & Sons, 2009.

Wicander, R.; Monroe, J. S. *Essentials of Geology*, 4th ed.; Thomson–Brooks/Cole: Belmont, CA, 2006.

Wilkinson, B.; Mcelroy, B.; Kesler, S.; Peters, S.; Rothman, E. Global Geologic Maps Are Tectonic Speedometers: Rates of Rock Cycling from Area-age Frequencies [Online] 2009. http://strata.geology.wisc.edu/reprints/Wilkinson_etal2009.pdf (Accessed Oct 1, 2017).

Williams, H.; Turner, F.; Gilbert, C. *Petrography*; W.H. Freeman: San Francisco, 1982.

Winkler, H. G. *Petrogenesis of Metamorphic Rocks*; Springer: Berlin, 1965.

Zhu, Y.; An, F.; Tan, J. Geochemistry of Hydrothermal Gold Deposits: A Review. *Geosci. Front.* **2011,** *2*, 367–374.

CHAPTER 3

Earth's Energy Balance and Climate

SANDEEP NARAYAN KUNDU*

*Department of Civil & Environmental Engineering,
National University of Singapore, Singapore*

E-mail: snkundu@gmail.com

ABSTRACT

Earth receives energy externally from solar insolation and internally from the radioactive processes in its core. Being a dynamic system, the various subsystems of Earth interactively exchange matter and energy. The energy and material exchanges are in balance which keeps our planet within the temperature ranges that support life forms. It is therefore essential to understand the various components of the Earth system and the interactive processes which are responsible to maintain Earth's energy balance. In this chapter, we shall learn about various forms of energy, their sources, the exchange process between the various Earth subsystems, and the potential risks to Earth's energy balance.

3.1 ENERGY AND EARTH

Our Earth is a dynamic system where the interaction of matter and energy is manifested in the natural Earth processes around us. All these natural activities derive energy from different sources primarily as internal heat from radioactive decay of material in the Earth's mantle and core and external solar irradiation. The internal heat drives crustal and solar energy drives weather and erosion systems. The sum of the different kinds of energy in a system is called the internal energy of the system. Energy is exchanged between material within the Earth's subsystems and between Earth and Space. This exchange is in the form of heat transfer and our

Earth needs to maintain a balance in its internal energy to sustain the current processes and life forms. The energy balance of the Earth, and therefore the climate system, can be considered in terms of the net imbalance between absorbed shortwave radiation and outgoing longwave radiation at the top of the atmosphere (Wake, 2014). In this chapter, we shall learn about various forms of energy, their sources, the exchange process between the various Earth subsystems, and the potential risks to Earth's energy balance.

Energy assumes various forms and can be described as either potential energy or kinetic energy, or a combination of both. Potential energy is the stored energy of a system and can be described as chemical energy, the energy which holds the elements together in molecules of materials; as nuclear energy, the energy within the atom's nucleus which is released by fission and fusion reactions; as mechanical energy, the energy that can be stored in physical configurations like the wound spring of a watch; and as gravitational energy, the energy defined by position with respect to distance from Earth's center. Kinetic energy is energy expressed in the movement of matter and includes radiant energy, energy which travels in form of electromagnetic radiation, for example, light; electrical energy, manifested in form of transfer of electrons or other charged particles, for example, lightening; thermal energy or heat which is due to vibrational motion of atoms and molecules in matter; sound energy is released through alternating compression and expansion of matter; and motion of objects, for example, flowing river and a rolling stone. One form of energy transforms into another form during natural processes like meteorite impact (mechanical energy to thermal energy); plate margin tectonics (thermal energy to chemical energy and backward in the processes of rock melting and recrystallization); and evapotranspiration (thermal energy into kinetic energy and then potential energy).

3.2   EARTH'S ENERGY BALANCE

The Earth system receives 5000 times more energy from external sources, primarily the Sun, than from within its interior. The sum total of incoming solar energy and outgoing radiative energy from Earth is commonly referred to as the Earth's energy budget. The energy budget should be maintained for Earth to be in a radiative equilibrium and this is essential to maintain a steady temperature.

Twenty-nine percent of the incoming solar energy is reflected back by our atmosphere and Earth's surface. Gases, clouds, and other suspended particles in our atmosphere block a part of this incident solar energy and the remainder is reflected back by snow and ice cover and reflective surface features. Of the remaining 71%, only 23% is absorbed water vapor, ozone, and other particulates in the atmosphere and the remainder by the geosphere and the hydrosphere. The absorbed energy excites and agitates the atoms and molecules of substances increase its temperature and at the same time, the excess of heat is radiated to the cooler objects in the surroundings. This radiated heat is proportional to the fourth power of the temperature of the substance and continues till the temperature differential with the surroundings is negligible. This is called radiative cooling, which regulated the temperature of Earth as the heated Earth radiates heat energy back into space.

As 71% of incoming solar radiation is taken in by Earth, it is essential that this amount is radiated back to space to maintain a stable average temperature. The atmosphere, the geosphere, and the hydrosphere are asymmetric in their absorption and radiation. The atmosphere reflects 59% of the total energy it receives of which it absorbs only 23% of incoming solar energy. The remainder is allowed to pass through to the Earth's surface of which half is absorbed by land and water and only 12% is reflected back. This indicates that most solar heating happens at the surface whilst the atmosphere is responsible for most of the radiative cooling. To understand how the Earth's climate balances the energy exchanges, it is necessary that the processes occurring at the surface of the Earth (where most solar heating takes place); at the edge of Earth's atmosphere (where sunlight enters the system); and within the atmosphere, are accounted for. For Earth to maintain a uniform temperature the net flux, that is, the incoming and outgoing energy, must be equal.

3.2.1   SURFACE ENERGY BALANCE

The heat exchange at the air/ground interface is determined by many complex processes making the energy balance at the Earth's surface extremely difficult to quantify and model (Beltrami et al., 2000). However, it had been approximately estimated that the ocean and land surfaces absorb 48% of incoming solar energy which leaves the ground into the atmosphere through three primary processes: evaporation, convection, and emission of thermal infrared energy.

Evaporation: Water molecules absorb incoming solar energy and form water vapor. This is called evaporation. The heat energy required to evaporate water is latent (potential energy) in water vapor which escapes into the atmosphere. When the water vapor molecules condense and return back to the ground through precipitation, this latent heat is released back to the surrounding atmosphere. As more than 70% of our Earth's surface is covered with water, evaporation is the most predominant process for the transfer of energy from land to atmosphere. It is responsible for 25% of the energy that is transferred back from the Earth's surface to the atmosphere. The exchanges of momentum, heat, and moisture between the atmosphere and the Earth's surface have a fundamental influence on the dynamics and thermodynamics of the atmosphere (Schmugge and Andre, 1991).

Convection: When fluids are heated, its density changes and differential heating on a fluid body creates regions which vary in density. This initiates of matter in the fluid as less dense fluid rise creating a vacuum underneath to which denser fluids move into. This circulation also transfers heat within the fluid and is called convection. Air in the atmosphere is in direct contact with warm land and water bodies. Air near surface receives heat through convection and moves vertically up into the atmosphere. As the atmosphere is warmer at ground level compared to the higher altitudes, there is constant circulation of air, a process that results in wind systems. The flow of wind, both laterally and vertically account for about 5% of heat transfer from the Earth's surface to the atmosphere. This exchange which uses air as a medium is called convection

Thermal radiation: Thermal radiation is responsible for the rebound of 17% of the energy received by the Earth's surface. This mode is different from evaporation and convection as the energy travels back into the atmosphere as infrared waves which do not involve matter as a medium. This is a result of competing fluxes between the Sun and Earth's surface. Sun's radiative energy is mostly in visible and near-infrared wavelengths whereas the Earth's radiative heat is confined to the infrared band.

3.2.2   ATMOSPHERIC ENERGY BALANCE

Like the energy flux at the Earth's surface needs to balance, the energy flux into the solid Earth and to the atmosphere must balance too. The atmosphere absorbs 23% energy from solar insolation and receives 30% more from Earth's surface processes like evaporation (25%) and convection (5%) making a total of 53% of all the energy received on Earth. Earth observation satellites have determined that the atmosphere radiates 59% energy back into space in form of thermal infrared rays. About 5 -6% of the thermal radiation from Earth's surface is trapped by the atmosphere (Fig. 3.1)

Atmospheric gases like water vapor, carbon dioxide, methane, and other trace gases allow sunlight to pass into Earth, but restrict the infrared thermal rays which are radiated back from Earth. This is natural green-house effect and has a great effect on Earth's surface temperature keeping it 30° warmer than it would be if the atmosphere was absent.

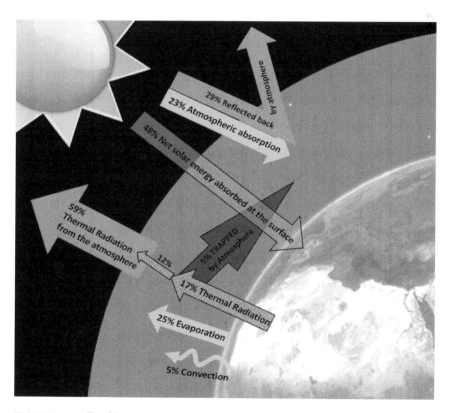

FIGURE 3.1   Earth's energy balance.

3.3   EARTH'S CLIMATE SYSTEM

Climate is the average weather of a region over a given period. These averages are worked out for a 30-year span, which as per The World Meteorological Organization is a standard time period for defining climate. Weather, on the other hand, is the instantaneous measure of temperature, precipitation, humidity, and near-surface wind direction and speed, which are primarily measured on an hourly basis by numerous weather stations installed around the globe. The climate of an area depends on latitude, proximity to the sea, vegetation cover, topography, and several other factors. Like the weather, the climate also varies at different time scales and geological evidence indicates that the Earth went through several ice ages in the past. The long-term mean of climate at a given place is very important since this drives the ecology and the environment useful for human living. Major climate zone is identified on Earth-based chiefly on latitude and other geographic factors. Variability in climate depends on the evolving interactivity between various components (Fig. 3.2) which collectively form the Climate System. Climate zones are classified based on a combination of climate variables (Stocker et al., 2013).

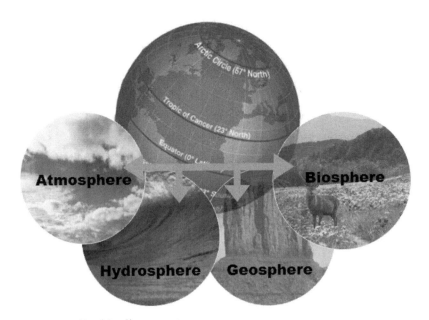

FIGURE 3.2   Earth's climate system.

3.3.1 CLIMATE ZONES

The three major climate zones on the Earth have been identified. These are the polar, the temperate, and the tropical zones and are represented by characteristic weather based on their geographical location, or latitude. The solar energy received per unit area for each of these climate zones depends on the angle of solar insolation. In equatorial regions, where the Sun's rays are more perpendicular and arrive at a shorter path, the maximum amount of heat is delivered. In polar regions, on the other hand, the Sun's rays arrive at an angle and travel longer to reach and therefore deliver lesser energy.

3.3.1.1 TROPICAL ZONE

In the regions between the equator and the subtropics (between 0° and max 40° latitudes), the proximity to the Sun as compared to the poles and the perpendicular positioning with respect to solar radiation at noon-time during almost the entire year makes this zone the hottest on Earth. High temperatures lead to accelerated convection, thermal radiation, and evaporation, generating frequent and dense which curtails incoming solar radiation to the surface (Fig. 3.3).

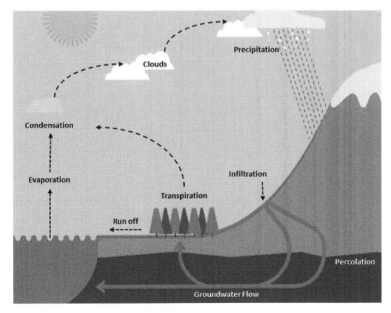

FIGURE 3.3 The hydrologic cycle.

3.3.1.2   TEMPERATE ZONE

In the temperate zone (between 40° and 60° latitudes), the solar radiation arrives at an angle which results in lesser heating than the tropics. In this zone seasons and length of day and night differ significantly in the course of a year which aids a more regular distribution of the precipitation round the year resulting in sustained vegetation cover for most times of the year.

3.3.1.3   POLAR ZONE

The polar areas (between 60° and 90° latitudes) receive the least heat from the Sun. This is because of the very flat angle of the Sun toward the ground and it also is the farthest as compared to the tropics and temperate zones. Because of the configuration of the Earth's rotation axis with the Sun, this zone has longer days and shorter nights during summer and the vice versa in winter. Freezing temperatures result in permafrost which melts only for a few months of summer during which sparse vegetation growth is supported.

3.3.2   CLIMATE SYSTEM COMPONENTS

Over geologic periods, the Earth climate is governed by the interplay of the carbon cycle involving the Earth atmosphere, terrestrial vegetation, oceans, sediments, and lithosphere, and the water cycle involving the atmosphere, rivers, and oceans. The recognition of multiple linkages between the Earth's physical environment and the biosphere is a discovery of the twentieth century, which led to the new concept of Earth System Science (Asrar et al., 2001). It has been established that the climate system of Earth is composed of the atmosphere, hydrosphere (includes cryosphere), and biosphere (includes anthroposphere). These interacting parts are our Earth's subsystems which are like four vast reservoirs in which energy and material are constantly being exchanged. Our Earth behaves as a closed system, except for a balanced energy exchange with outer space. Energy reaches Earth in abundance in form of solar radiation and also leaves the system in form of longer wavelength infrared radiation whereas matter exchanges between Earth and outer space are near to nonexistent (except for meteorites).

Satellites play a central role in assessing climate change because they can provide a consistent global view, important data, and an understanding of change (Lewis et al., 2010). Advances in the Earth observation from satellites and ground station measurements have added on to our basic understanding of the Earth system and its cycles, such as clouds, land surface, water and ecosystems, water and energy cycles, oceans, glaciers and polar ice, and solid Earth. Thorough understanding of these cycles at various temporal scales has improved natural hazard prediction, natural resource management, and anthropogenic environmental impacts.

3.3.2.1   THE GEOSPHERE

The geosphere comprises the solid Earth principally composed of the crust, mantle, and inner core. The upper part of the crust which consists of rocks and loose unconsolidated soil is the most happening place in terms of energy and matter interactions with other spheres of the Earth. This part is where external energy (solar radiation) meets the internal heat (radioactive heat from mantle) and where surface processes like weathering, erosion, and sedimentation occur. Crustal processes like plate tectonics (which create oceans and mountains) and surface processes (weathering, erosion, and sedimentation) happen here including many of the dynamic and hazardous events like landslides, earthquakes, and volcanic eruptions which threaten our modern society.

3.3.2.2   THE HYDROSPHERE

The hydrosphere comprises all of Earth's water stored in liquid and solid states in oceans, inland water bodies, groundwater, and snow. A subsystem of the hydrosphere consisting of perennially frozen parts of water is being increasingly important and is referred to as the cryosphere. Water also exists in form of vapor in our atmosphere alongside other gases, but from a system point of view, this is not considered as a part of the hydrosphere. Water through its wide occurrence in different physical forms connects all the subsystems of Earth. This connectivity is in form of natural processes which allow the exchange of water between these subsystems. Through the hydrologic cycle, this exchange forms an important life-sustaining service to our environment. Water is a solvent and accommodates several materials within itself and the hydrological cycle ensures that pure water is always available on Earth.

3.3.2.3   THE ATMOSPHERE

The atmosphere is a gaseous envelope around the geosphere and is commonly referred to as "air," The atmosphere extends up to 10,000 km above ground with a density of gases decreasing gradually as we move further away from Earth. Our atmosphere contains 78% nitrogen, 21% oxygen, 1% argon, and 0.04% carbon dioxide by volume in addition to a small number of trace gases. The atmosphere also contains 0.04% water is a form of vapor, the proportion of which increases to 1% over large water bodies. Apart from gases our atmosphere also holds suspended particulate matter. The atmosphere regulates the incoming heat from the Sun and protects our Earth from overheating. Atmospheric oxygen supports life on Earth. Natural processes exchange heat and gases between geosphere, biosphere, and hydrosphere leading to variations in the composition, density, pressure, and quality of the atmosphere. Photosynthesis is a critical process which replenishes oxygen and sequestrates carbon dioxide thereby sustaining life and thermos regulating the environment, respectively.

3.3.2.4   THE BIOSPHERE

The biosphere is the sum total of all life forms on Earth and also includes undecomposed organic remains of life. The biosphere is unique to Earth as compared with other planets of the solar system and beyond. The essential differences between life forms and nonliving matter are in the processes of metabolism, reproduction, growth, and evolution and each of these processes involve energy and material exchanges with other spheres of Earth. Metabolism refers to the chemical reactions within an organism which produces the energy for other life processes and requires food which is an intake of matter. Autotrophs produce their own food from inorganic chemicals through photosynthesis whereas heterotrophs feed on other organic matter for metabolism. The biosphere exists as an ecosystem which if different for different environments. Early Earth had no biosphere as evidenced by the geologic records and the evolving planet and its subsystems gradually led to existence of early life forms which evolved to the present day complex forms of life around us. Humans, which for the anthroposphere, and their activities are significantly impacting our environment, ecosystem, and other subsystems. This had led to the evolution of anthroposphere as a major subsystem of our Earth system which is critical to climate change and global warming.

3.3.3   INTERACTIVE CYCLES

Energy and material exchanges between the four reservoirs define climate through several coupling mechanism commonly known as the biogeo-chemical cycles (Bengtsson and Hammer, 2001). The prime ones are the water cycle, the carbon cycle, and the rock cycle.

3.3.3.1   THE HYDROLOGIC CYCLE

The hydrologic cycle governs the movement of water on the Earth involving the geosphere and the atmosphere and utilizes energy from solar irradiation. The bulk of Earth's water is contained in the oceans, which is about 30,000 times the amount present in the atmosphere and inland water bodies combined. Approximately 70% of the Earth's surface is covered with water and the average depth of water in oceans is about 4 km. This leaves a very small fraction of total water on Earth available and suitable for human use which occurs on land. Ice caps and glaciers account for 3% of total water and water contained in the pored of rocks or groundwater is a mere 1%. Both these occurrences constitute the terrestrial fresh water supply. The water transfer between these various locations and forms are aided by evaporation, transpiration, precipitation, surface run off, and percolation.

The atmosphere (0.001%) and the biosphere (0.0001%) contain little water as compared to water as snow and in oceans but play a vital role. Evaporation and precipitation are vital processes through which energy is exchanged and balance is achieved between these climate system components. Respiration also involves the exchange of water between the biosphere and the atmosphere and all these processes generate fluctuations in the amount of the greenhouse gases (GHGs) driving weather and climate.

Within the geosphere itself, about 30,000–40,000 km^3 of water is exchanged between the surface of the continents and the oceans through evaporation–precipitation-runoff processes which shape the topography of the continents. Surface heating from solar insolation drives evaporation transforming liquid water in oceans to vapor in the atmosphere. Condensation forms clouds which under the effect of gravitational pull results in precipitation generating runoff into river systems, eroding loose Earth, cutting gorges and canyons, and transporting sediments in suspension and solution which eventually land in the oceans. The transfer of water from the oceans to the atmosphere is simultaneous with the transfer of tremendous amounts of thermal energy to the atmosphere. Wind systems also add this energy circulation through convective processes. In permafrost regions, precipitation is in the form of snow which accumulates to form glaciers which move slowly downslope under gravity eroding the bedrock and transporting sediments to form landforms like moraines and till. Glaciers end up either in the sea forming icebergs or melt below the snowline feeding melt water into rivers.

There are several subsystems of the hydrologic system which operate in their own cycles. Examples are river systems, glacial systems, groundwater systems which not only shape the landscape but also influence weather and climate systems through energy exchange. Sun is the primary energy source which drives these surface processes where energy is circulated and constantly redistributed on the Earth.

3.3.3.2   THE CARBON CYCLE

Carbon is crucial to both energy and life. All living things are made of carbon and hydrocarbons are the primary composition of fossil fuels. Carbon is an important constituent of all the climate components and it is circulated between them in different chemical forms and this constitutes the carbon cycle (Fig. 3.4).

In the atmosphere, carbon exists as called carbon dioxide which is taken in by plants to form carbohydrates, lipids, and other organic compounds through the process of photosynthesis. Photosynthesis is the process which exchanges carbon between the atmosphere and the biosphere. Some plants are consumed by animals as a part of the food chain moving carbon from plants to animals. Both plants and animals die and decompose to release carbon back into the atmosphere. Sedimentary processes also take a part of this dead organic matter and store it in the rocks of the geosphere. The carbon is taken in as organic matter in these rocks transform into fossil fuels like coal and oil over millions of years. When humans burn fossil fuels, most of this trapped carbon is released back into the atmosphere as carbon dioxide.

Carbon dioxide is a greenhouse gas and helps the atmosphere trap the thermal heat emanating from the Earth's surface. As long as the concentration of carbon dioxide is constant in the atmosphere the Earth's energy budget is in balance. Too much of carbon dioxide will lead to the Earth getting warmer impacting climate, too less would lead to a frozen world.

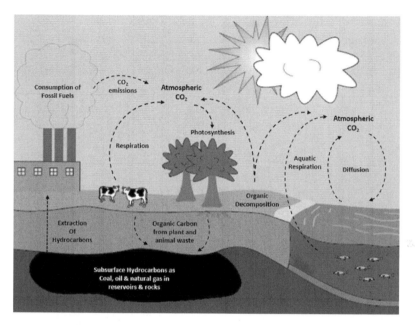

FIGURE 3.4 The carbon cycle.

3.3.3.3 THE ROCK CYCLE

Rocks are the prime constituents of the geosphere. There are three primary types of rocks, namely igneous, sedimentary, and metamorphic rocks. Time-consuming transitions through geologic time transform one type of rock into another which is conceptually known as the rock cycle (Fig. 3.5). An igneous rock such as basalt may break down and dissolve when exposed to the atmosphere or melt as it is subducted under a continent.

The driving forces of the rock cycle are plate tectonics and the hydrological cycle which does not allow the rocks to be in equilibrium because of pressure and temperature changes and the processes of weathering and erosion. The abundance of water on Earth plays a very important role for the rock cycle, mainly because of the water driven sedimentary processes involving weathering and erosion. Precipitation intakes carbon dioxide from the atmosphere making the water acidic which chemically erodes minerals in rocks which are more reactive. Most minerals found in igneous and metamorphic rocks are unstable under near surface and atmospheric conditions and are more susceptible to chemical erosion. The runoff from precipitation carries away the ions dissolved in solution and the broken down products of weathering into rivers which end up in the oceans basins. The dissolved and suspended sediments are then deposited in these basins and eventually compact forming sedimentary rocks. The rock cycle illustrates how the three rock types are related to each other, and how processes change from one type to another over time. Its involvement in carbon exchange through sedimentary accumulation of organic matter makes it an important biogeochemical cycle.

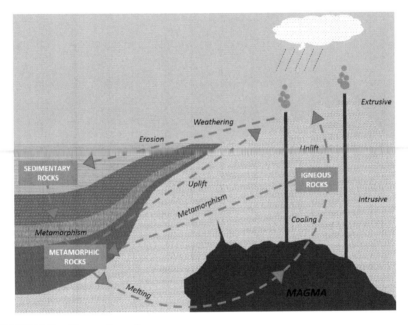

FIGURE 3.5   The rock cycle.

3.4   CLIMATE CHANGE

Understanding how and why climate has changed in the past is important to identify climate drivers. The main source of decadal variability in the tropics and subtropics is the natural forcing and internal variability of surface temperatures (Terray, 2012). Since state of the art observational record is very short, we have to rely on proxy data to estimate the past climate. Proxy data like tree rings, ice cores, fossil pollen, ocean sediments, corals, and so on help climatologists reconstruct the past climate. It has been established that Earth has undergone several warm and cold phases is the past. The cold phases were witnessed by large glacial cover on Earth known as ice ages with inter glacial warm phases in between. Several drivers of climate change have since been which are also cyclic in natures. These are internal and external drivers which impact climate with varying severity and in different repetitive cycles.

3.4.1   EXTERNAL CLIMATE DRIVERS

External drivers of climate (Fig. 3.6) are primarily because of the planetary configuration of Earth and the fluctuations of solar energy. The eccentricity of Earth's orbit and the tilt of its axis vary systematically causing cyclic changes in solar insolation at any region. This produces climate modulations through repetitive temperature variation cycles. These variations are termed as Milankovitch cycles, and are understood to be the reason behind the various ice ages and warmer climates in between.

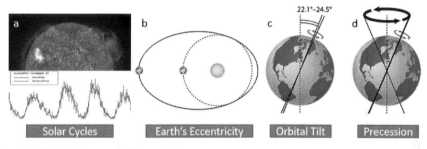

FIGURE 3.6   External climate drivers.

3.4.1.1   SOLAR CYCLES

The only external source of energy received by Earth is from the Sun, which is located 149 million km away. The Sun's temperature cyclically every 11 years as observed from the study of sunspots which are dark splotches on the Sun's surface. Sunspots have been a crucial marker for understanding the mechanisms that rule the Sun's interior. Over the course of a solar cycle, the sunspots are found to migrate progressively toward the lower latitudes.

3.4.1.2   EARTH'S ECCENTRICITY

Earth revolves around the Sun in an elliptical orbit, eccentricity of which changes from a near circular one to elliptical on a 100,000-year cycle. The changes in eccentricity lead to variations in range of temperature variations on Earth.

3.4.1.3   ORBITAL TILT

The Earth's spin axis is not vertical to the elliptical plane of its orbit. It is at an angle which varies between 22° and 24°. This impacts the angle of solar insolation resulting variations in the intensity of the solar heat received at a location.

3.4.1.4   PRECESSION OF ROTATION

The Earth wobbles on its axis as it spins like a rotating top. This wobble (or precession), changes the tilt of the axis toward and away from the Sun over a period of 19,000–23,000 years. These minor changes in geometrical configuration alter the amount of sunlight incident on each hemisphere of Earth impacting severity of seasons on its year-long voyage around the Sun.

These external drivers are believed to be the reason behind the time interval of cooler temperatures between the mid-14th and the mid-19th century, known as the "Little Ice Age." The Little Ice Age brought an end to an unusually warm "Medieval climate optimum" resulting in bitterly cold winters to many parts of the world which are well documented in Europe and North America. During this phase, glaciers in the Swiss Alps expanded in area engulfing farms and entire villages. The River Thames and the canals and rivers of the Netherlands often froze over during the winter, and people skated and even held fairs on the ice. Famines and epidemics were more frequent impacting human life.

3.4.2   INTERNAL CLIMATE DRIVERS

3.4.2.1   GEODYNAMICS

Our Earth's crust consists of continental and oceanic plates which are in constant motion resulting in collision and subduction. This crustal motion is called plate tectonics and is responsible for mountain building, seafloor spreading, crustal thinning, and volcanism which are all transient and slow processes which shape the Earth's morphology and the proportional distribution of oceans and continents. This distribution is important because it determines the routing of global current systems and the oceanic heat exchange intensity between equatorial and Polar Regions. The movement of the plates modulates the global climate on very long time scales by altering wind patterns, ocean currents, and the accumulation of snow and ice. Under certain configurations of landmass and ocean distribution, the heat exchange can be severely impeded. For example, if all the landmasses aggregate at the poles, the warm ocean currents may not be effective which shall result in glaciated continents and therefore an ice age occurs. In contrast, if all the continents aggregate around the equatorial region, the Polar Regions shall have excellent good oceanic heat exchange leading to uniform global temperatures.

3.4.2.2   ALBEDO

Albedo is the proportion of solar energy which is reflected from the Earth's surface back into space. In other words, it is the reflectivity of the Earth's surface. Land cover plays an important role in albedo with ice/snow cover having a very high albedo. In contrast, water is less reflective. Earth devoid of snow cover therefore shall lead to more absorption of solar energy and lead to warming of climate.

3.4.2.3   GREENHOUSE EFFECT

Of the 17% energy in form of thermal infrared radiation emanating out of the Earth's surface, 5–6% is trapped by the atmosphere. Gases in the atmosphere such as carbon dioxide aid trapping of thermal infrared energy whereas the same atmospheric gases allow all other forms of energy to pass through it creating a greenhouse effect. The greenhouse effect is beneficial as it keeps our Earth warm and habitable. However, an abundance of gases like carbon dioxide, water vapor, methane, nitrous oxide, and ozone, commonly known as GHGs, will lead to overheating of Earth impacting climate. With anthropogenic emissions of GHGs being the highest in history, its impacts on climate systems is already being felt through increased frequency of extreme climate and natural disasters. This is emerging as the prime driver of climate change today as evidenced by widespread impacts on human and natural systems.

3.4.2.4   VOLCANISM

Volcanism results from geodynamic processes and volcanic events spray materials to the Earth's atmosphere hindering the incoming solar irradiation. These materials stay suspended in the atmosphere for a long time and gradually reach the surface through sever bouts of precipitation. Past volcanic events have triggered omissions of seasons in certain years. Two hundred years ago, when Mount Tambora erupted it shot ash miles into the sky which spread as far across the Northern Hemisphere and left the Earth in the midst of a year without a summer.

3.4.2.5   PATTERNS AND TRENDS

There is an overwhelming interest in climate change as anthropogenic activities are increases GHGs leading to global warming, impacts of which has the potential to disrupt human life like the Little Ice Age of the 17th century. Satellite-based observations of Earth's surface temperatures have yielded strong correlation with atmospheric CO_2 concentrations, establishing CO_2 as the prime driver of climate change. Vegetation dynamics can provide valuable information about global warming, phenological change, crop status, land degradation, and desertification (Liu et al., 2015). The investigation of the variability of the land temperature is also of great importance (Varotsos and Alexopoulos, 1979). Modern evidences from satellite observation, combined with proxy evidence from Earth's geological records have helped climate reconstructions of the past. Recent works have evidences that since the preindustrial era, anthropogenic GHG emissions have increased to alarming levels. Largely driven by economic and population growth, current estimates atmospheric concentrations of carbon dioxide, methane and nitrous oxide are now unprecedented since the last 800,000 years. Human use of land for agriculture and the receding snow in the Arctic have also significantly reduced albedo.

3.4.2.6   IPCC SCENARIO-BASED TRENDS

Predicting how future global warming will contribute to climate change needs many factors to be accounted for. The amount of future greenhouse gas emissions is a key variable and therefore socioeconomic aggregations from all the nations are required which includes considerations on developments in technology, changes in energy generation and land use, global and regional economic circumstances, and population growth.

To complement and compare different climate research, Intergovernmental Panel on Climate Change, in 2013–2014, presented a standard set of scenarios which ensures that starting conditions, historical data and projections are employed consistently across the various branches of climate science. These scenarios are called the representative concentration pathways (RCPs) of which four are commonly used to depict future scenarios (Fig. 3.7). RCP 2.6 presents a case for maintaining similar levels of surface temperature which would lead to climate stability. If emissions are not maintained at this level and reach the RCP 8.5 scenario then global temperature could rise by 3° at least and this amount is enough to invite extreme climatic variations which would be disruptive to humanity.

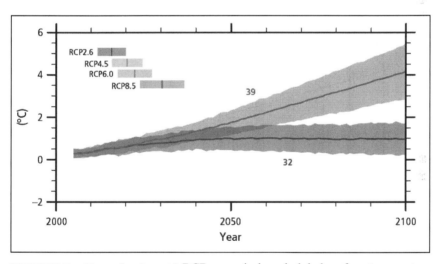

FIGURE 3.7 **(See color insert.)** RCP scenario-based global surface temperature change predictions.

3.5 CONCLUSION

Rigorous analysis of a wide variety of data collected with varying spatial and temporal scales reveal that most of the observed global warming in the recent past does not seem to bear a natural footprint. What has changed in the last 50 years is postindustrial anthropogenic activity which has changed the Earth's energy balance through accelerated greenhouse emissions and land surface degradation. It is therefore necessary that we develop means to discern the human influence on climate by developing models which accommodate many natural variations of climate factors which influence land and sea surface temperatures, precipitation, and other aspects at various scales.

Natural events which impact climate change can be both periodic and sporadic. Periodic evidences are the Sun cycles and the eccentricity changes whereas sporadic ones are atmospheric aerosols resulting from events volcanic eruptions or peat forest fires. Different influences on climate lead to different patterns observed in climate records. Observed atmospheric temperature trends provide a fingerprint which is more associated with gradual increase in CO_2 levels in the past 30 years that to natural events like the Sun cycles. With increased earth observation from space, scientists routinely test the dynamic parameters to study and explain the patterns of climate change and update the factors influencing climate system. We still have very little high frequency records for the past, which is an impediment to predict future climate and with each day our understanding about Earth's energy balance and climate system is being advanced.

KEYWORDS

- **hydrologic cycle**
- **carbon cycle**
- **rock cycle**
- **greenhouse effect**
- **solar cycle**

REFERENCES

Asrar, G.; Kaye, J. A.; Morel, P. NASA Research Strategy for Earth System Science: Climate Component. *Bull. Am. Meteorol. Soc.* **2001,** *82* (7), 1309–1329.

Beltrami, H.; Wang, J.; Bras, R. L. Energy Balance at the Earth's Surface: Heat Flux History in Eastern Canada. *Geophys. Res. Lett.* **2000,** *27* (20), 3385–3388. DOI: 10.1029/2000GL008483.

Bengtsson, L.; Hammer, C. U. *Geosphere–Biosphere Interactions and Climate*; Cambridge University Press: Cambridge, 2001.

Lewis, J. A.; Ladislaw, S. O.; Zheng, D. E. *Earth Observation for Climate Change*; Center for Strategic and International Studies, 2010.

Liu, Y.; Li, Y.; Li, S.; Motesharrei, S. Spatial and Temporal Patterns of Global NDVI Trends: Correlations with Climate and Human Factors. *Remote Sens.* **2015,** *7* (10), 13233–13250. DOI: 10.3390/rs71013233.

Schmugge, T. J.; André, J. *Land Surface Evaporation: Measurement and Parameterization*; Springer: New York, NY, 1991.

Stocker, T. F.; Qin, D.; Plattner, G. K.; Tignor, M.; Allen, S. K.; Boschung, J.; Nauels, A.; Xia, Y.; Bex, B.; Midgley, B. M. *Climate Change 2013: The Physical Science Basis. Contribution of Working Group I to the Fifth Assessment Report of the Intergovernmental Panel on Climate Change*; IPCC, 2013.

Terray, L.; Evidence for Multiple Drivers of North Atlantic Multi-decadal Climate Variability. *Geophys. Res. Lett.* **2012,** *39* (19). DOI: 10.1029/2012GL053046.

Varotsos, P.; Alexopoulos, K. Possibility of the Enthalpy of a Schottky Defect Decreasing with Increasing Temperature. *J. Phys. C* **1979,** *12,* L761–L764.

Wake, B. Radiative Forcing: Earth's Energy Balance. *Nat. Clim. Change* **2014,** *4* (9), 758–758. DOI: 10.1038/nclimate2364.

Zhang, X.; Yan, X.; Chen, Z. Geographic Distribution of Global Climate Zones Under Future Scenarios. *Int. J. Climatol.* **2017,** *37* (12), 4327–4334. DOI: 10.1002/joc.5089.

CHAPTER 4

Mass Extinctions on Earth

SANGHAMITRA PRADHAN[1], SHREERUP GOSWAMI[1,*], and
SANDEEP NARAYAN KUNDU[2]

[1]*Department of Earth Sciences, Sambalpur University, Burla 768019,
Odisha, India*

[2]*Department of Civil & Environmental Engineering,
National University of Singapore, Singapore*

Corresponding author. E-mail:goswamishreerup@gmail.com

ABSTRACT

In the history of 4.6 billion years, Earth has experienced many mysterious and unexplained events of obliteration, which demarcates the boundaries among geological periods. Fossil records are fundamental to the interpretation of these past geological events. Earth was punctuated by five major extinction events. Each event eliminated most of the flourishing species and raised the curtain for new species after. These five major mass extinctions, which occurred during Late Ordovician, Late Devonian, Permian–Triassic transition, Triassic–Jurassic transition, and Cretaceous–Tertiary transition, are often referred to as the five largest Phanerozoic mass extinctions. Climate change and global oceanic circulation played a significant role in these "Big Five" events that wiped out more than half of the pervasive biota. Global ecosystem was disturbed and ruptured many times due to phenomenon like marine transgressions and regressions, global warming and climate change, ocean anoxia, asteroid impact, and intense volcanic activities in the total Earth's history. This chapter focuses on "Mass Extinction," which is perhaps the most perplexing event on Earth, and emphasizes on its relative causes and consequences.

4.1   INTRODUCTION

Earth has formed 4.6 billion years ago. The geological history of Earth has witnessed many major and minor events based on the geological time scale (GTS). The geological timescale presents a chronological scheme, which integrates stratigraphic records classified based on the geochronological sequence of events. This timescale is extensively used by earth scientists for describing the timing and relationship among these events, which present a vivid understanding of our Earth's evolution in an integrated manner. It consists of standard stratigraphic divisions based on rock or strata and their evolutionary sequence calibrated in years (Harland et al., 1982). The divisions show the positions in the form of chronostratigraphic units and time as geochronological units which are correlated at various temporal and spatial scales which, can be hinged to have a global scale understanding of the sequence of events. The GTS is hierarchical with Eon as the major divisions which are further subdivided into Era and then into Epochs or Series (Fig. 4.1).

In the 18th century, attempts were made to standardize the GTS to expand its applicability anywhere on Earth. Early attempts divided the rocks into four chronologic units, namely, primary, secondary, tertiary, and quaternary with each unit corresponding to a specific period in Earth's history. Pioneered by geoscientists like William Smith and Alexandre Brogniart, more precise differentiation of the rock within each unit were attempted using fossil records contained in them. Fossils helped correlate distant strata, separated by continental distances and detailed studies in the 19th century produced a sequence of stratigraphic units, which we still use today.

The names in the geologic time scale reflect the dominance of British geologists. Based on stratigraphic sequences in the wales, names like "Cambrian," "Ordovician," and "Silurian" were included. "Devonian" was derived from the county Devon, which was used for rocks strata found in the place. Some other names like "Triassic" was names based on the three distinct layers—red beds, chalk, and black shales—which were found throughout Germany and Northwest Europe. "Cretaceous" (derived from Latin term creta for chalk) was defined using the strata in Paris basin where extensive beds of chalk deposited by marine invertebrate shells. The first GTS was published in 1841 based on fossils found in each era where terms like "Paleozoic" (meaning old life), "Mesozoic" (meaning middle life), and Cenozoic (new life) were used

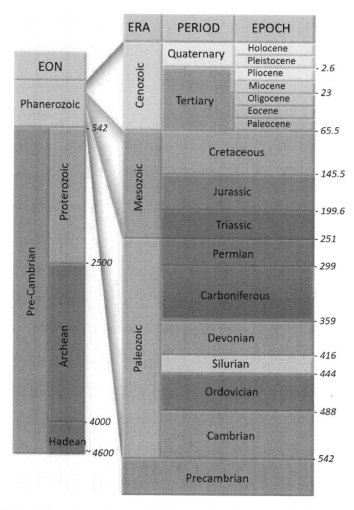

FIGURE 4.1 The geological time scale. Ages are in million years (myr).

Source: Sourced from USGS. (https://pubs.usgs.gov/of/2008/1206/images/tbl3_1.jpg)

based on fossil content in the strata. This emphasized the importance of fossils and life forms on Earth's history as an important parameter for defining boundaries between past times. Fossil contents are governed by evolutionary trends and extinctions on Earth which also are important indicators of climatic changes in Earth's past. Understanding mass extinctions presents an important mode to discourse past changes and predicting our future.

For nongeoscientists, the concept of geologic time is an enigma as most are used to deal time in increments of minutes, hours, days, months, and years and not millions or billions of years. In the context of Earth's 4.6-billion-year history, geologists often consider an event which is 100 million years old as recent. An appreciation of geologic time scale is important for understanding extinctions on Earth as geologic processes and evolution of life on Earth are so gradual processes that vast spans of time are needed before significant extinctions occur.

The geologic time scale divides the vast 4.6-billion-year history of Earth into eons, eras, periods, and epochs. Currently, we are in the Holocene epoch of the Quaternary period which belongs to the Cenozoic era of Phanerozoic eon. Past dates for the time divisions in the GTS have been established using relative and absolute dating techniques. The Precambrian phase, consisting of eons like Hadean, Archean, and Proterozoic, accounts for more than 88% of the GTS. The GTS is not periodic as same designations do not necessarily extend for the same number of years. For instance, the Cambrian period comprises 54 million years whereas the Silurian is only 28 million years. This is because the GTS is based on changing character of life forms on Earth through time as evidenced by rock records. Later eons have more subdivisions than earlier ones as recent rock records possess diverse fossils, evolution trails of which demark the boundaries effectively. Older rock records have sparse fossils and therefore demarcation of the periods and epochs are uncertain. Also, Precambrian rocks are more likely to have been metamorphosed destroying most evidence necessary for such classification. Biodiversity of life forms was low in older times in the evolutionary history of the Earth.

Eons are further divided into several Eras. Geological Era is a division of geological time spanning for several hundred million years. Archean is the earliest Era began about 4600 myr. Paleozoic Era commenced about 570 myr, Mesozoic at 250 myr and Cenozoic, the most recent one initiated about 65 myr. The geological eras further subdivided into several periods. The major subdivisions of the Paleozoic Era are Cambrian, Ordovician, Silurian, Devonian, Carboniferous, and Permian. Mesozoic Era is subdivided into Triassic, Jurassic, and Cretaceous subdivisions. The recent Era holds Quaternary and Tertiary periods as major subdivisions. The Cambrian is the oldest started on about 570 myr while the most recent period is Quaternary initiated about 2.5 myr. Epochs are the divisions of Period covering about tens of millions of years. Now we are living in the most topical epoch, that is, Holocene epoch, set in motion about 1.8 myr.

4.2   FOSSILS AND GEOLOGIC HISTORY

One of the most important inclusions in sediments and stratified rocks are fossils which are essentially traces of prehistoric life which are preserved in the rocks containing them. Fossils are fundamental in interpreting geologic past and constitute a science called paleontology which incorporated understandings of geology and biology in order to provide insights into the evolutionary aspects of life forms on Earth over the vast expanse of our Earth's history. The nature of life forms provides evidence of the past climatic and environmental conditions based on which depositional environments of the rocks and the predominant processes can be inferred. Presence or absence of fossil assemblages in rocks near and far helps in their correlation which established geological concepts like continental ages and past ice ages.

4.2.1   PRESERVATION OF LIFEFORMS

Fossils are found in many different forms and can be classified into various types (Fig. 4.2). Most fossils are organic remains which are resistant to weathering like teeth, bones, and shells. Other soft remains like flesh and other material are rarely preserved with only examples like that of the woolly mammoth which was preserved in permafrost regions. Remains of an organism decompose differently, based on the climate of the region where they proliferate. Some fossils are petrified remains, where the voids created by the decomposed life form are filled with naturally occurring minerals preserving the microscopic details of the organic structure at times. Moulds are created when the material is dissolved by water which is later casted by mineral deposits which assumed the external form of the organism. Another mode is carbonization where leaves of plants and woody matter are buried along with fine-grained sediments. An example of such is fern found in coal seams.

In many cases, the remains are not preserved at all but the existence of the life form is evident from its activities around it dwelling area. These are tracts and trails of life forms which are preserved in sedimentary rocks. Tracks are footprints of animal movement and can include evidence of

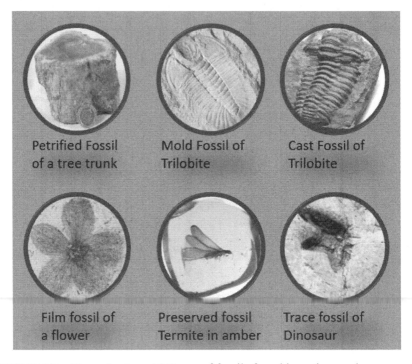

FIGURE 4.2   (See color insert.) Types of fossils found in rock records.

the existence of small organisms like earthworms to the large dinosaurs. Burrowing animals leave tubular structures in soft sediments which are often filled with later sediments of a different nature preserving these borrow structures in the lithified rock. Dungs of organisms are also preserved and also constitute a part of this category called trace fossils.

4.2.2   INDEX FOSSILS

Based on observations and findings of many geologists, an important principle in historical geology—the theory of faunal succession—was formulated. This theory applied the Darwinian theory of evolution to fossils as they document the evolution of life through time. Fossil organisms, as per this theory—succeed one another in a definite and determinable order and based on the stratigraphy time zones can be identified which help in determining the subdivisions in the GTS.

Through fossils, paleontologists have identified ages in the GTS which belonged to fishes, coal swamps, reptiles, and mammals based on their abundance and succession in sedimentary records. Fossils, therefore, are recognized as time indicators and their extinction can always be related to a prehistoric event marking the end of a time division and beginning of another. To facilitate this process of division, a class of fossil organisms which came into existence and got extinct rapidly serve as a clock for geologic time. Such fossils are called index fossils and are very important in determining the lower subdivisions of the GTS.

4.3 MASS EXTINCTIONS

In the life succession of organisms, extinction is a phase, which witnesses the wiping out of the species from the planet. Mass extinction is identified by a rapid decrease in global biodiversity or a sharp change in abundance of biota in the intervals, which demarcates the boundary between geological periods. Such extinctions are evidenced as a global scale and therefore is a widespread event, which is often associated with major climate change events which effect a rapid decrease in global biodiversity or a sharp change in abundance of biota.

Mass extinctions distinguish the boundary between geological periods, which denote extinction of a set of species and evolution of another set of species. Understanding the process of extinction and the factors responsible for it is very much crucial to understand the evolution of biosphere, environment, and climate change. In the history of Earth's 4.6 billion years, it has gone through five major mass extinctions, which are well documented in the literature (Fig. 4.3). These occurred during late Ordovician, late Devonian, Permian–Triassic transition, Triassic–Jurassic transition, and Cretaceous-Tertiary transition and are often referred to as the five largest Phanerozoic mass extinctions.

Most of the species that have ever lived on the Earth surface got extinct and many new species have been evolved over passing Eras. Present day is key to the past, the analysis of fossil records indicate the extinction rates in biota, which have not been constant through the time, also help in delineating the paleoclimate, paleoenvironment, and the approximate causes of the mass extinction events. The event mass extinction is a sharp decline in biodiversity demarcates the boundary between two geological periods. In the flow of this event, global marine biodiversity as well as

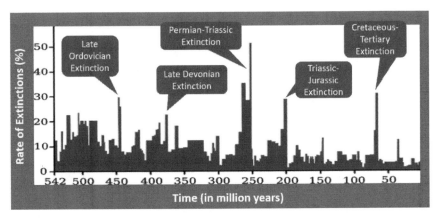

FIGURE 4.3   Major mass extinctions on earth.

Source: Modified from Wikipedia Commons.
(https://en.wikipedia.org/wiki/File:Extinction_intensity.svg)

terrestrial flora and fauna have been devastated for many times during the Earth history resulting in collapse and reordering of total marine as well as the land ecosystem. Understanding the evolutionary role of mass extinction what drives global biodiversity requires detailed knowledge of postextinction recoveries that control diversity at a variety of geographical and temporal scale.

4.4   MAJOR EPISODES OF EXTINCTIONS

4.4.1   LATE ORDOVICIAN

In the Era of Phanerozoic, the first ever mass extinction is the late Ordovician mass extinction (LOME) during which 85% of marine biota were eliminated (Sepkoski, 1996). The LOME is unambiguously linked with climate changes (Fig. 4.4) which triggered extinction in two pulses (Finnegan et al., 2002).

The primary pulse was triggered by the initial glaciation stage when the sea level declined and resulted in a harsh climate in low and mid-latitudes (Sheehan, 2001). This was followed by rapid growth of ice sheets in southern poles on Gondwanaland (Brenchley et al., 2001; Kaljo et al., 2008) expansion of which led to substantial cooling of tropical oceans (Vandenbroucke et al., 2010; Finnegan et al., 2011). Such cooling influenced the flow of deep ocean currents, which used to bring in nutrients from the temperate zones to the tropics. In this pulse, which changed the global carbon cycle (Brenchley et al., 2003; Jones et al., 2011; Kump et al., 1999) and witnessed huge drop in eustatic levels (Sheehan, 1988; Kaljo et al., 2008; Finnegan et al., 2011) lead to mass extinction of several marine species. The stable isotopic composition of marine carbon in the rocks indicated global environmental changes like the growth of ice sheets and drop in ocean temperature, which drove major changes in the global carbon cycle (Fig. 4.4).

The surviving fauna post this pulse adapted the new ecological and environmental setting. However, toward the end of the glaciations which brought in a reverse change in climate lead to the rise of sea level and changes in ocean circulation triggered the second pulse of extinction. Suddenly the glaciations period ended, as there was a drastic change in climate, which leads to the rise in sea level and changes in ocean circulation, which initiated the second pulse of extinction.

In the Ordovician, period many tectonic plates were dispersed at the equator. During Late Ordovician, the huge Gondwana landmass wandered to the South. The Ordovician atmosphere had a higher amount of CO_2 than today and it together with other greenhouse gases compensated for reduced solar insolation which was about 5% lesser than at present. The Paleozoic evolutionary fauna (EF) were mostly marine and were less diverse than their modern counterparts. Communities were primarily epifaunal which dwelled on the surface of rocks in marine conditions unlike the infaunal (burrowing) communities of today. The structures of communities were simpler with less number of predators than modern EF. During the Late Ordovician event, the Cambrian EF got extinct and those who survived were deep offshore dwellers.

According to Peters (2006), the common cause or the sedimentary record for the LOME is explained by two hypotheses. One is the Eustatic Common Cause hypothesis, which identifies the Gondwana glaciations as the prime driver of the mass extinctions, which lowered the eustatic levels affecting the shallow marine fauna most of which then became extinct (Newell, 1967; Johnson, 1974). The other hypothesis the Common Climate Cause, which postulates the cooling of temperatures that drives coastal regressions and habitat losses which are accountable for the high extinction rate (Stanley, 1984, 1988).

It took several years to recover from this major event and peculiarly the new fauna, which evolved had similar ecological patterns to the ones which

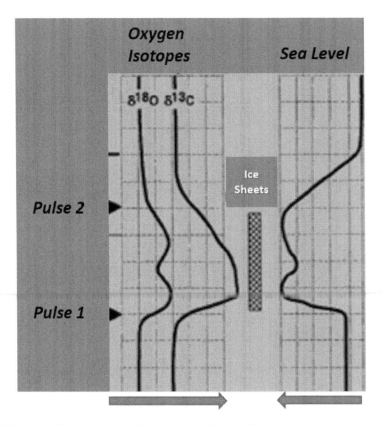

FIGURE 4.4   The two pulses of extinction during late Ordovician and associated with isotopic oxygen changes and eustasy.

Source: Modified from Sheehan (2001).

become extinct. Both climatic and sea level changes were deemed to be the main causes of the Late Ordovician Extinction. The cooling of the global climate was definitely the reason behind the decline of biodiversity. The accompanied coastal regression eliminated many of the habitats of endemic communities (Cocks and Rickards, 1988; Sheehan, 1988). Productivity fell down as nutrient levels declined in the ocean waters. In the mid-Ordovician period, prior to the approach of Gondwana to the South Pole, the oceans were warm and salinity stable (Railsback et al., 1990) which was disturbed in the Late Ordovician, due to thermohaline circulations which exist in the present day (Berry and Wilde 1978; Wilde et al., 1990). Upwelling of nutrient-rich water to the surface stimulated the primary product. The sudden change of glaciations brings in rapid eustatic and isotopic changes (Fig. 4.4) which explain the climatic conditions responsible for the LOMEs.

According to the Sepkoski (1996), the taxonomic loss in the LOME involved the perishing of 26% families and 49% genera of the then biodiversity. It was reported that about 85% of all species were extinct (Jablonski, 1991). Despite being severe in the loss of taxa, this episode was less significant in terms of ecological consequences as the subsequent biota which evolved after were largely drawn from the ones which had survived the extinction event.

Brachiopods were the most common member among benthic faunas during Ordovician, whose population declined significantly at the end of this mass extinction event. Many groups like Athyrids, Pentamerids, and Spiriferids survived this extinction and later flourished. The population of some clades such as Orthids, Strophomenids declined only slightly. Inarticulate forms like Lingulata and Craniata suffered the most significant decline (Harper and Rong, 1995). Some fauna survived into the Silurian and others became extinct at the end of glaciations as the sea level rose (Rong and Harper 1988). Some of Corals and Stromatoperoids got extinct during the first pulse of Ordovician extinction. In the temperate realm Rugosa corals dominated, whereas tabulate corals became less diverse and stromatoporoids became too sparse. Post this episode new solitary rugosa genera evolved and some new tabulate corals were introduced as the sea level rose after the postglaciations (Elias and Young, 1998; Young and Elias, 1995). The second pulse of the extinction had little effect that the corals simply took advantage of the newly available habitat. The extinction of Echinoderms was concentrated in the first pulse of the Ordovician extinction event. Cystoids were very common in the Ordovician but were replaced by Crinoids after Late Ordovician. Some genera of Crinoids which survived the climatic disaster has a very similar morphology to the ones which got extinct (Foote, 1999). Many minor Echinoderms disappeared across the extinction event. 13% of Bryozoans declined in this event with two dominant groups, the Cryptostomata and the Trepostomata got extinct depleting their generic diversity (McKinney and Jackson, 1989). 80% of Conodont species also disappeared (Sweet, 1990) and a third of the flourishing Ostracod families declined across the Ordovician extinction event (Copeland, 1981). Among the Cephlopods, Nautiloids experienced severe destruction (Crick, 1990). The highly endemic Late Ordovician faunas were replaced by most cosmopolitan faunas in the Phanerozoic. The Late Ordovician glaciations spanned only half a million years (Brenchley et al., 1994) which was short but the effect was severe as it drove a mass of species to their extinction.

4.4.2   LATE DEVONIAN EVENT

Post the Late Ordovician episode, another major extinction, known as the Late Devonian Extinction occurred on Earth. This event was a very unusual mass extinction as it was reflected differently at different places and also the declination of biota was not concentrated into a single sharp pulse. Of the many pulses identified, the five main ones spread over 1.0–1.5 myr, which were responsible for eliminating about 82% of marine biota (Jablonski, 1991; McGhee, 1996). The timing and duration of Late Devonian mass extinction are still being debated owing to different interpretations and varying geographic evidence. According to McGhee (1996), several extinction pulses were grouped under the Late Devonian extinction event of which the lower Kellwasser event was the first ever extinction pulse. This was followed by three extinction events collectively called as the Upper Kellwasser events which occurred in late Frasnian. Named after the Kellwasser limestone units of Germany (Walliser, 1996), all these three extinction pulses are believed to have occurred during the last 100–300 thousand years of Frasnian stage (Walliser, 1996).

The Late Devonian extinction was spread over 25 myr with the highest rate of decline in biota peaking in the Frasnian (Sepkoski, 1982). Some consider this extinction as a prolonged depletion of marine biota spanning 20–25 myr which punctuated by 8–10 extinction events (Thomas et al., 2000). Among these 8–10 events, two were significant extinction events with the first one being the Kellwasser event which occurred at the Frannian–Famennian boundary. The other event was the Hangeberg event, believed to have happened close to the Devonian-Carboniferous boundary.

Some researchers like House (2001) are of a different view as they consider the Famennian stage to have staged the highest family extinction of marine biota which was followed by the Givetian and then by the Frasnian. Based on the rate of extinction, the Givetian stage is the severest where 14.2 family extinctions occurred per million years. The second and third severe stages experienced elimination of 11.2 families and 6.8 families, respectively.

During the Late Devonian, the Gondwana supercontinent covered many parts of the Southern hemisphere and the continents were arranged differently (Mc Kerrow et al., 2000). The Devonian diversity included land plants, which were very similar to Ordovician mosses, liverworts, and lichens all of which had developed roots, seeds, and vascular system, for the transportation of water and nutrients. Such evolutionary changes post the Late Ordovician event aided the survival of plants in environments away from swampy habitats. Such adaption created large forests in the highlands in the Devonian period. This period also witnessed the evolution of the first jawless fishes in Earth history. Nevertheless, the entire ecosystem collapsed mostly because of an anoxic ocean, decrease in atmospheric CO_2, global cooling and increased asteroid impacts, all of which led to the mass extinction in Late Devonian.

The invertebrates of Silurian period continued to evolve and flourish in the Devonian which proliferated in the shallow and deep marine habitats. Primitive fishes in both fresh and marine water habitats evolved in the Devonian period, which is also commonly known as the "Age of Fish."

The most probable causes of the Late Devonian extinction event are climatic in nature (McGhee, 1989). Global warming (Thompson and Newton, 1988), sea level changes, dysoxia, or anoxia of Devonian seas and oceans (Joachimski and Buggisch, 1993) were the prime attributes. The anoxic state of oceans was evidenced by the global distribution of black shales in association with Kellwasser limestone units. The anoxic water spread into shallow water regions and caused a severe decline in shallow water diversity. High latitude ecosystem, deepwater marine faunas, and terrestrial ecosystem were slightly affected by anoxia. The anoxic ocean waters were more extensive in the tropics than in the temperate region. The mass deaths and subsequent burial of organic matter and carbonates in the marine environment depleted the atmospheric CO_2 level and also contributed to the deposition of black shales. The loss of CO_2, a greenhouse gas, was the prime driver for the global cooling for a shorter span (McGhee, 1996).

It is also believed that photosynthetic sequestration of CO_2 by extensive forests also contributed to the rapid decline of atmospheric CO_2 levels, which could have led to drop in temperatures. The evolution of the vascular system of plants and trees in Devonian accelerated rock weathering producing chemically broken matter, which produced nutrients that were transported to the oceans (Algeo and Scheckler, 1998).

Increased nutrients to the oceans could have caused eutrophication causing anoxia.

A short-term transgression regression was experienced at the Frasnian–Famennian boundary (Buggisch, 1991) which indicated switching of global climate from Frasnian greenhouse to a Famennian ice house which led to a regressive phase toward the end of Famennian into Carboniferous. This change impacted temperate biota which then got extinct.

Several tremendous mass extinction events in Devonian caused crises in global biodiversity in which both marine and terrestrial ecosystem collapsed. The Devonian reef ecosystem, which was the most geographically widespread ecological unit that had ever exist in the Earth's history (10 times larger than the reef ecosystems of today), was severely impacted. In the Frasnian–Famennian event the tropical, as well as peri-reefal ecosystems, were ruptured. These included stromatoporoids, rugosa, and tabulate corals. The decline was so remarkable that the major elements of reef ecosystem could not recover for the remainder of Paleozoic Era. Other victims were brachiopods, conodonts, ammonoids, jawless fish, placoderms, trilobites, and benthic foraminifera. The Devonian period was marked as "Golden Age" for brachiopods, as their populations dominated the Paleozoic seas. Almost 75% of brachiopods genera including two super families, atrypacea and pentameracea were eliminated in the Late Devonian extinction event. Most of the extinct species particularly the biconvex brachiopods could not recover for the remainder of their evolutionary history (McGhee, 1988). Apart from the benthic ecosystem, the other impacted ecosystems were the planktons and nektons, 90% of which were diminished (McGhee, 1988). A few families of trilobites which included scutelluidae, odontopleuridae, and tropidocoryphidae and some ammonoids like manticeras) and concodont genera became extinct. Only three conodont species out of 47 could survive (Sorauf and Pedder, 1986; Sorauf, 1989). Post this event the Trilobites which evolved had smaller eyes indicating their adaptation to benthic habitats or turbid waters.

4.4.3 PERMIAN–TRIASSIC EVENT

250 myr ago, Earth witnessed the largest of all extinctions which defined the demarcation between the Permian and the Triassic. This extinction introduced a new era, the Mesozoic era, in the geological time table. A series of short yet sever pulses contributed to the most devastating

extinction event in the Earth's record (Erwin, 1993; 1994; Benton, 1995) which had the highest ecological impact and biodiversity loss and equally impacted the marine as well as a terrestrial ecosystem (McGhee et al., 2004). T Near about 90–96% of marine biota and approximately 70% terrestrial ecosystem were annihilated (Benton, 2005; Sahney and Benton, 2008; Carl and Lee, 2012). The Permian–Triassic event is the only known extinction event which involved the extinction of insects (Labandeira and Sepkoski, 1993; Sole and Newman, 2003). Recovery of biodiversity took a significantly longer period than other extinction events (Benton, 2005). This period, unlike the previously discussed events, witnessed intense short and prolific pulses of continental volcanic eruptions which severely impacted biota.

Though the triggers behind the Permian–Triassic extinction still debated, many researchers have cited evidence of extensive volcanism and bolide impact and its domino effects which influenced climatic change, ocean water anoxia, methane release, hypercapnia, hydrogen sulfide poising, and global warming. Triggering factors have been grouped under two categories, first of which are catastrophic (fissure volcanism and bolide impacts) followed by a series of environmental triggers (ocean anorexia, sulfide poisoning, and global warming-related sea level changes).

Based on evidence from North Western Australia and Antarctica, bolide (asteroid or comet) impact appears to be the most plausible reason for the predicament of Permian biomass (Becker et al., 2004). Fullerenes, which are molecules of carbon in the form of a hollow sphere or ellipsoidal tubes, have been detected in deposits associated with various bolide events with the Earth. These are normally found from shock-produced breccia and contain extraterrestrial gases, shocked quartz, meteoritic metallic elements, and nano-phase iron oxides. The mineral grains are rich in iron, nickel, silicon (Railback et al., 1998; Backer et al., 2001; Kaiho et al., 2001; Backer et al., 2004). Fullerenes were abundantly found in Permian rocks which support the bolide theory of mass extinction.

The huge eruption of continental flood basalt (Fig. 4.5) from fissure volcanism has also been found accountable for biological catastrophes (Wignall, 2001; Courtillot and Renne, 2003; Racki and Wignall, 2005). Fissure volcanism, as in the case of Deccan traps of Western Indian craton, also ejects poisonous gases like halogens, halides into the atmosphere increasing the amount of CO_2 thereby causing hypercapnia. The released SO_2 and CO_2 cause acid rain contributing to the toxicity of the environment which the Permian biota could not tolerate.

FIGURE 4.5   Permian flood basalts of Siberia.
Source: The Geological Society, UK. (https://www.geolsoc.org.uk/flood_basalts_1)

Rapid climatic changes also played a major role in the extinction of Permian biota. Hypercapnia-associated global warming and increased aridity were likely to have caused paleofires which are evidenced by the distinct charcoal layers in Permian rocks (Darcy and Norman, 2011). Warming of climate also could have lowered the solubility of O_2 in oceans causing anorexia in the photic zones (Grice et al., 2005). The widespread occurrence of black shales in the lowermost Triassic shelf is likely to have been caused by the death of Permian planktons due to anoxia. Hypercapnia, H_2S poisoning, the addition of SO_2, the release of halogen and halides from the flood basalt volcanism could also have added to the already existing woes.

The Permian–Triassic event is the largest extinction event where marine invertebrates, land plants and vertebrates suffered the most. Among the impacted marine invertebrates, articulate brachiopods, ceratitida, and crinoids became almost extinct and their diversity declined rapidly (Villier and Korn, 2004; Leighton and Schneider, 2008). Moderately impacted marine invertebrates included trilobites, rugose corals, tabulate corals and echinoderms. Among the aquatic vertebrates, jawed fishes, numerous families of insects, reptiles and amphibians were impacted. This extinction event was so vigorous that most of the impacted genera could not recover. Some of the groups who survived in thin populations suffered a long-term size reduction relative compared to times before this event. These groups along with some other leftover biotas experienced post-event syndrome popularly known as the Lilliput effect, used to describe a decrease in body size in animals which have survived a major extinction.

4.4.4 TRIASSIC–JURASSIC EVENT

The Triassic–Jurassic event marks the boundary between the Triassic and Jurassic period occurring at the end of the Triassic period. This major extinction event of the Phanerozoic Eon affected the terrestrial and marine lives significantly. About 76% of species both in land and water and approximately 20% of all taxonomic families got extinct in this event. This event marked the evolution of reptiles and dinosaurs become dominant land animals on Earth. The severity of this extinction was low (fourth among the five mass extinctions) in which water diversity reduced by 21% impacting major marine groups which included 8 families of sponges, 13 families of gastropods, 12 families of brachiopods, some bivalves, cephalopods, and various marine reptiles. On land, this event eliminated most of the insects (35 families lost), fresh-water bony fishes, and the conodonts.

Some authors suggested that there were three pulses of extinction in this Triassic event, each of which was caused by multiple reasons most likely of gradual sea level changes, bolide impact and related temperature changes, and fissure volcanism. A major asteroid impact has been implicated in the end of Triassic mass extinction event (Olsen, et al., 1987; Olsen and Cornet, 1988) with evidence of shocked quartz lenses from Triassic–Jurassic layer supported the bolide impact, which might be a triggering element of the Triassic extinction event. The Triassic climate was much warmer than today and dense vegetation extended beyond higher latitudes than they are at present times. The poles were free from ice caps and climatic conditions were hot and arid. The atmosphere had CO_2 levels thrice high than present-day levels which enhanced the greenhouse effect. This coupled with the large scale of flood basalt volcanism at end Triassic impacted the extinction pulses. Intense volcanism also pumped in a lot of CO_2 into the atmosphere causing temperature rise. This was in addition to the introduction of toxic gases like SO_2, H_2S, halides, and halogens from the volcanic eruptions which impacted the Triassic livelihoods. The changing climate resulted in the rapid fluctuation of sea level throughout the Triassic–Jurassic period impacting coastal habitats and thereby reducing biota diversity. Both regression and ocean water anoxia are established causes of reductions of habitat extends of seas and have the potential for driving mass extinctions (Hallam, 1989).

4.4.5 CRETACEOUS-TERTIARY EVENT

65.5 million years ago a global scale and abruptness of the major biotic turnover were recognized by Paleontologists at the Cretaceous-Tertiary boundary. The extinction event is also called K-T boundary event which witnessed a widespread extinction episode where approximately 80% of all species of Cretaceous fauna became extinct. "K" stands for Kreide, the German term for chalk and "T" for Tertiary, which represents the time between Paleogene and Neogene periods. The K–T extinction was marked by the abolition of many biotas which became important rudiments of the Mesozoic Era and accounted for nearly all of the dinosaurs and many marine invertebrates. Based on the severity, the K–T extinction ranks third among the five major extinction episodes. Marine, as well as terrestrial biota (including insects, mammals, birds, flowering land plants, fishes, corals, and mollusks), went on to diversify tremendously in Early Cretaceous. At this K–T boundary, 40% of marine invertebrate and 75% terrestrial vertebrate species were not found from the fossil records. Well-known groups such as dinosaurs (non-avian) like pterosaurs, mosasaurs, and ammonites (and many marine planktons) species got eliminated from the Earth's history.

The K–T extinction event is the most recent mass extinction in the Earth's history and is believed to have been caused by catastrophic events. The catastrophic events include gigantic asteroid impact and/or increased volcanism. The event coincided with the Chicxulub large asteroid impact and with the flood basalt volcanism in the western Indian craton. Such catastrophes brought in large forest fires and atmospheric aerosols which reduced solar insolation for long periods consequences of which include extinction of 60% of flora and all the dinosaurs.

Voluminous eruptions of magma and the associated release of sulfur and carbon dioxide may have caused rigorous environmental effects with the sulfur rapidly transforming into the sunlight-absorbing sulfur aerosols that cooled the Earth for many years (Pierazzo et al., 2003).

These catastrophes did not impact the deep ocean temperatures because of oceanic huge thermal mass, which most likely contributed to a rapid recovery of the global climate aftermath of the event. The released sulfur also produced acid rain but it may not have been sufficient to alter the acidity of the ocean significantly as it would have has on terrestrial water bodies, which is why ocean biota was less impacted then terrestrial biota. Major ecosystems collapsed on the Earth and the biotas who survived adapted into smaller forms and flourished later.

4.5 CONCLUSION

Mass extinctions are annihilation events and are noxious. Despite the fact, these events present opportunities for newer life forms to emerge, evolve and flourish. Mass extinctions are biotic crisis caused by environmental stimulus presented by natural processes like fissure volcanism and flood basalts, asteroid impact collisions, climatic changes like glaciation and sea level changes, ocean anoxia and changes in the atmospheric composition of gases. Such changes push the environmental conditions to the extreme, which challenges the wellbeing of several life forms which are stressed and eventually are wiped out.

Mass extinctions also trigger organic evolution. The hasty loss of biota and major compositional transformations point to the then global paleoecological cataclysm at the mass extinction periods (McElwain and Punyasena, 2007). The catastrophes altered gene expression in biota (Comai, 2005) which contribute to hybrid vitality which produces resilient forms in adaptation with new and emerging conditions (Rieseberg et al., 2003, 2007; Hegarty et al., 2008). Partitioning of ancestral expression patterns in response to environmental stresses produces subfunctional duplicated genes (Force et al., 1999) which then undergoes genetic evolution. Differential expression of these duplicated genes leads to new hybrid forms, which are more adaptable in the emerging environmental conditions.

The evidencing of mass extinctions in our Earth's past reflects the even changing environmental and climatic conditions on Earth and indicates that such mass extinctions are a reality and can happen in the future pointing to the sixth (next) mass extinction. With anthropogenic influences trending ahead of natural drivers, the future appears challenging for global biota, the diversity of which is at stake. Recent evidence of global temperature rise and its correlation with increasing carbon levels in our atmosphere indicate an increasing human footprint on the changing environment. Evidence of environmental hazards are ion the rise, climate shifts have impacts food production and industrial production has impacted balanced land use. All these have a drastic influence on our environment challenging its sustainability and have all the ingredients to trigger another mass extinction. Increased industrialization, rapid urbanization, deforestation and crop monocultures, overfishing, use of nuclear weaponry, and accelerated fossil fuel usage, all have the potential to influence deglaciation and eustatic changes endangering habitats of flora and fauna. Should these continue unabated, the sixth mass extinction would certainly set in during the next centuries.

KEYWORDS

- fossil
- climate change
- ocean anoxia
- eustatic changes
- bolide impact
- volcanism

REFERENCES

Algeo, T. J.; Scheckler S. E.; Maynard J. B. *Effects of the Middle to Late Devonian Spread of Vascular Land Plants on Weathering Regimes, Marine Biota, and Global Climate*; Columbia University Press: New York, 2000; pp 213–236.

Algeo, T. J.; Scheckler, S. E. Terrestrial-marine Teleconnections in the Devonian: Links Between the Evolution of Land Plants, Weathering Processes, and Marine Anoxic Events. *Philos. Trans. R. Soc. B: Biol. Sci.* **1998,** *353* (1365), 113–130.

Becker, L.; Poreda, R. J.; Basu, A. R.; Pope, K. O.; Harrison, T. M.; Nicholson, C.; Iasky, R. Bedout: A Possible End-permian Impact Crater Offshore of Northwestern Australia. *Science* **2004,** *304,* 1469–1476.

Benton, M. J. Diversification and Extinction in the History of life. *Science* **1995,** *268,* 52–58.

Berry, W. B. N.; Wilde, P. Progressive Ventilation of the Oceans: An Explanation for the Distribution of the Lower Paleozoic Black Shales. *Am. J. Sci.* **1978,** *278,* 27–75.

Brenchley, P. J.; Carden, G. A. F.; Hints, L.; Kaljo, D.; Marshall, J. D.; Martma, T.; Midla, T.; Nalvak, J.; High-resolution Stable Isotope Stratigraphy of Upper Ordovician Sequences: Constraints on the Timing of Bioevents and Environmental Changes Associated with Mass Extinction and Glaciation. *Geol. Soc. Am. Bull.* **2003,** *115,* 89–104.

Brenchley, P. J.; Marshall, J. D.; Carden, G. A. F.; Robertson, D. B. R.; Long, D. G. F.; Meidla, T.; Hints, L.; Anderson, T. F. Bathymetric and Isotopic Evidence for a Short-lived Late Ordovician Glaciation in a Greenhouse Period. *Geology* **1994**, *22*, 295–298.

Brenchley, P. J.; Marshall, J. D.; Underwood, C. J. Do all Mass Extinctions Represent an Ecological Crisis? Evidence from the Late Ordovician. *Geol. J.* **2001**, *36*, 329–340.

Buggisch, W. The Global Frasnian-Famennian "Kellwasser Event." *Geol. Rundsch.* **1991**, *80*, 49–72.

Cocks, L. R. M.; Rickards, R. B. Eds. Global Analysis of the Ordovician-Silurian Boundary. *Bull. Br. Mus. Nat. Hist. Geol.* **1988**, *43*, 1–94.

Colbert, E. H. Mesozoic Tetrapod Extinctions: A Review. In *Dynamics of Extinction*; Elliott, D. K., Ed.; John Wiley and Sons: New York, 1986; pp 49–62.

Comai, L. The Advantages and Disadvantages of Being Polyploid. *Nat. Rev. Genet.* **2005**, *6*, 836–846.

Copeland, M. J. Latest Ordovician and Silurian Ostracode Faunas from Anticosti Island Qu´ebec. In *Field Meeting, Sub-commission on Silurian Stratigraphy;* Ordovician-Silurian Boundary Working Group: Anticosti -Gaspé, Quebec, 1981; pp 185–195.

Courtillot, V. E.; Renne, P. R. On the Ages of Flood Basalt Events. *C.R. Geosci.* **2003**, *335*, 113–140.

Crick, R. E. Cambrian-Devonian Biogeography of Nautiloid Cephalopods. *Geol. Soc. Mem.* **1990**, *12*, 147–161.

Elias, R. J. Young, G. A. Coral Diversity, Ecology and Provincial Structure During a Time of Crisis: The Latest Ordovician and Earliest Silurian Edgewood Province in Laurentia. *Lethaia* **1998**, *13*, 98–112.

Erwin, D. H. *The Great Paleozoic Crisis, Life and Death in the Permian*; Columbia University Press: New York, 1993; p 327.

Erwin, D. H. The Permo-Triassic Extinction. *Nature* **1994,** *367*, 231–236.

Fagerstrom, J. A. *The Evolution of Reef Communities*; John Wiley and Sons: New York, 1987; p 600.

Finnegan, S.; Bergmann, K.; Eiler, J. M.; Jones, D. S.; Fike, D. A. Eisenman, I.; Hughes, N. C.; Tripati, A. K.; Fischer, W. W. The Magnitude and Duration of Late Ordovician-Early Silurian Glaciation. *Science* **2011**, *331*, 903–906.

Foote, M. Morphological Diversity in the Evolutionary Radiation of Paleozoic and Post-paleozoic Crinoids. *Paleobiol. Mem.* **1999**, *25*, 1–115.

Force, A.; Lynch, M.; Pickett, F. B.; Amores, A.; Yan, Y. L.; Postlethwait, J. Preservation of Duplicate Genes by Complementary, Degenerative Mutations. *Genetics* **1999**, *151*, 1531–1545.

Grice, K.; Cao, C.; Love, G. D.; Böttcher, M. E.; Twitchett, R. J.; Grosjean, E.; Summons, R. E.; Turgeon, S. C.; Dunning, W.; Jin, Y. Photic Zone Euxinia During the Permian–Triassic Superanoxic Event. *Science* **2005**, *307*, 706–709.

Hallam, A. The End-Triassic Bivalve Extinction Event. *Palaeogeogr. Palaeoclimatol. Palaeoecol.* **1981**, *35*, 1–44.

Hallam, A. The Case for Sea-level Change as a Dominant Casual Factor in Mass Extinction of Marine Invertebrates. *Philos. Trans. R. Soc. Lond. Ser. B* **1989**, *325*, 637–655.

Harper, D. A. T.; Rong, J. Patterns of Change in the Brachiopod Faunas Through the Ordovician-Silurian Interface. *Mod. Geol.* **1995**, *20*, 83–100.

Harper, D. A. T.; Rong, J.; Zhan, R. Late Ordovician Development of Deep-water Brachiopod Faunas. *Acta Univ. Carol. Geol.* **1999**, *43*, 351–353.

Hegarty, M. J.; Barker, G. L.; Brennan, A. C.; Edwards, K. J.; Abbott, R. J.; Hiscock, S. J. Changes to Gene Expression Associated with Hybrid Speciation in Plants: Further Insights from Transcriptomic Studies in Senecio. *Philos. Trans. R. Soc. Lond. Ser. B* **2008**, *363*, 3055–3069.

House, M. R. Chronostratigraphic Framework for the Devonian of the Old Red Sandstone. In *New Perspectives on the Old Red Sandstone*; Friend, P. F., Williams, B. P. J., Eds.; Geological Society, Special Publications, 2001; Vol. 180, pp 23–27.

Jablonski, D. Extinctions: A Paleontological Perspective. *Science* **1991**, *253*, 754–757.

Joachimski, M.; Buggisch, W. Anoxic Events in the Late Frasnian: Causes of the Frasnian-Famennian Faunal Crisis. *Geology* **1993**, *21*, 675–678.

Johnson, J. G. Extinction of Perched Faunas. *Geology* **1974**, *2*, 479–482.

Jones, D. S.; Fike, D. A.; Fennegan, S.; Fischer, W. W.; Schrag, D. P.; McCay, D. Terminal Ordovician Carbon Isotope Stratigraphy and Glacioeustatic Sea-level Change Across Anticosti Island (Quebec, Canada). *Geol. Soc. Am. Bull.* **2011**, *123*, 1645–1664.

Kaljo, D.; Hints, L.; Mannick, P.; Nolvak, J. The Succession of Hirnantian Events Based on Data from Baltica: Brachiopods, Chitinozoans, Conodonts, and Carbon Isotopes. *Est. J. Earth Sci.* **2008**, *57*, 197–218.

Kump, L. R.; Arthura, M. A.; Patzkowskyb, M. E.; Gibbsa, M. T.; Pinkbusb, D. S.; Sheehanc, P. M. A Weathering Hypothesis for Glaciation at High Atmospheric pCO (2) During the Late Ordovician. *Palaeogeogr. Palaeocl. Palaeoeco.* **1999**, *152*, 173–187.

McElwain, J. C.; Punyasena, S. W. Mass Extinction Events and the Plant Fossil Record. *Trends Ecol. Evol.* **2007**, *22*, 548–557.

McGhee, G. R. The Multiple Impacts Hypothesis for Mass Extinction: A Comparison of the Late Devonian and the Late Eocene. *Palaeogeogr. Palaeoclimatol. Palaeoecol.* **2001**, *176*, 47–58.

McGhee, G. R. Jr. The Frasnian-Famennian Extinction Event. In *Mass extinction: Processes and Evidence Donovan*; S. K., Ed.; Columbia University Press: New York, 1989; pp 133–151.

McGhee, G. R. Jr. *The Late Devonian Mass Extinction: The Frasnian/Famennian Crisis*; Columbia University Press: New York, 1996; p 303.

McGhee, G. R.; Sheehan, P. M.; Bottjer, D. J.; Droser, M. L. Ecological Ranking of Phanerozoic Biodiversity Crises: Ecological and Taxonomic Severities are Decoupled. *Palaeogeogr. Palaeoclimatol. Palaeoecol.* **2004**, *211*, 289–297.

McKerrow, W. S.; Mac Niocaill, C.; Dewey, J. F. The Caledonian Orogeny Redefined. *J. Geol. Soc.* **2000**, *157*, 1149–1154.

McKinney, F. K.; Jackson, J. B. C. *Bryozoan Evolution*; Unwin Hyman: Boston, 1989; p 238.

Newell, N. D. *Revolutions in the History of Life*; Geological Society of America Special Publication, 1967; Vol. 89, pp 63–91.

Olsen, P. E.; Cornet, B. *The Triassic-Jurassic Boundary in Eastern North America*; Houston, Texas, Lunar and Planetary Institute Contribution, 1988; Vol. 673, pp 135–136.

Olsen, P. E.; Shubin, N. H.; Anders, M. H. New Early Jurassic Tetrapod Assemblages Constrain Triassic–Jurassic Tetrapod Extinction Event. *Science* **1987**, *237*, 1025–1029.

Paul, C. R. C. *Extinction and Survival in the Echinoderms*; Extinction and Survival in the Fossil Record, Systematics Association Special, 1988; Vol. 34, pp 155–170.

Peters, S. E. Genus Extinction, Origination, and the Durations of Sedimentary Hiatuses. *Paleobiology* **2006,** *32,* 387–407.

Pierazzo, E.; Hahmann, A. N.; Sloan L. C. Chicxulub and Climate: Radiative Perturbations of Impact-produced S-bearing Gases. *Astrobiology* **2003,** *3,* 99.

Racki, G. Wignall, P. B. Late Permian Double-phased Mass Extinctions and Volcanism: An Oceanographic Perspective. In *Understanding Late Devonian and Permian–Triassic Biotic and Climatic Events: Towards an Integrated Approach*; Over, D. L., Morrow, J. R., Wignall, P. B., Eds.; Elsevier, 2005; pp 263–297.

Railsback, L. B.; Ackerley, S. C.; Anderson, T. F.; Cisne, J. L. Paleontological and Isotope Evidence for Warm Saline Deep Waters in Ordovician Oceans. *Nature* **1990,** *343,* 156–59.

Raup, D.; Sepkoski, J. Mass Extinctions in the Marine Fossil Record. *Science* **1982,** *215,* 1501–1503.

Rieseberg, L. H.; Raymond, O.; Rosenthal, D. M.; Lai, Z.; Livingstone, K.; Nakazato, T.; Durphy, J. L. Schwarzbach, A. E.; Donovan, L. A.; Lexer, C. Major Ecological Transitions in Wild Sunflowers Facilitated by Hybridization. *Science* **2003,** *301,* 1211–1216.

Rieseberg, L. H.; Kim, S. C.; Randell, R. A.; Whithey, K. D.; Gross, B. L.; Lexer, C.; Clay, K. Hybridization and the Colonization of Novel Habitats by Annual Sunflowers. *Genetica* **2007,** *129,* 149–165.

Rong. J.; Harper, D. A. T. A Global Synthesis of the Latest Ordovician Hirnantian Brachiopod Faunas. *R. Soc. Edinb. Trans. Earth Sci.* **1988,** *79,* 383–402.

Self, S, and Rampino, M. Flood basalts, mantle plumes and mass extinctions. https://www.geolsoc.org.uk/flood_basalts_1. Accessed April 1, 2019.

Sepkoski, J. J. *A Compendium of Fossil Marine Families*; Milwaukee Public Museum Contributions in Biology and Geology, 1982; Vol. 52, p 124.

Sepkoski, J. J. Patterns of Phanerozoic Extinction: A Perspective from Global Data Bases. In *Global Events and Event Stratigraphy in the Phanerozoic*; Walliser, O. H., Ed.; Springer: Berlin, 1996; pp 35–51.

Sheehan, P. M. Brachiopods from the Jerrestad Mudstone (Early Ashgillian Ordovician) from a Boring in Southern Sweden. *Geol. Palaeontol.* **1973,** *7,* 59–76

Sheehan, P. M. *Late Ordovician Events and the Terminal Ordovician Extinction*; New Mexico Bureau of Mines and Mineral Resources Memoirs, 1988; Vol. 44, pp 405–415.

Sheehan, P. M. The Late Ordovician Mass Extinction. *Annu. Rev. Earth Planet. Sci.* **2001,** *29,* 331–364.

Sorauf, J. *Rugosa and the Frasnian-Famennian Extinction Event: A Progress Report*; Memoir of the Association of Australasian Palaeontologists, 1989; Vol. 8, pp 327–338.

Sorauf, J.; Pedder, A. Late Devonian Rugose Corals and the Frasnian-Famennian Crisis. *Can. J. Earth Sci.* **1986,** *23* (9), 1265–1287.

Stanley, G. D. The History of early Mesozoic Reef Communities: A Three-step Process. *Palaios* **1988,** *3,* 170–183.

Stanley, S. M. Temperature and Biotic Crises in the Marine Realm. *Geology* **1984,** *12,* 205–208.

Stanley, S. M. Climatic Cooling and Mass Extinction of Paleozoic Reef Communities. *Palaios* **1988,** *3,* 228–232.

Stanley, S. M. Paleozoic Mass Extinctions-shared Patterns Suggest Global Cooling as a Common Cause. *Am. J. Sci.* **1988,** *288,* 334–352.

Sweet, W. C. *The Conodonta: Morphology, Taxonomy, Paleoecology, and Evolutionary History of a Long-extinct Animal Phylum*; Clarendon: Oxford, UK, 1990; p 212.

Thompson, J. B.; Newton, C. R. Late Devonian Mass Extinction: Episodic Climatic Cooling or Warming. In *Devonian of the World Volume III*; McMillan, N. J., Embry, A. F., Glass, D. J., Eds.; Canadian Society of Petroleum Geologists Memoir, 1988; Vol. 14, pp 29–34.

Tozer, E. T. Triassic Ammonoidea; Classification, Evolution, and Relationship with Permian and Jurassic Forms. In *The Ammonoidea: Systems Assessment Special*; House, M. R., Senior, J. R., Eds.; Academic Press: London, 1981; Vol. 18, pp 66–100.

Twitchett, R. J. Incompleteness of the Permian-Triassic Fossil Record: A Consequence of Productivity Decline. *Geol. J.* **2001,** *36,* 341–353.

Vandenbroucke, T. R. A.; Howard, A. A.; Williams, M.; Paris, F.; Zalasiewicz, J. A.; Sabbe, K.; Nolvak, J.; Challands, T. J.; Verniers, J.; Servais, T. Polar Front Shift and Atmospheric CO_2 During the Glacial Maximum of the Early Paleozoic Icehouse. *Proc. Natl. Acad. Sci. USA* **2010,** *107,* 14983–14986.

Walliser, O. H. Global Events in the Devonian and Carboniferous. In *Global Events and Event Stratigraphy in the Phanerozoic*; Walliser, O. H., Ed.; Springer: Berlin, 1996; pp 225–250.

Ward, P. D.; Botha, J.; Buick, R.; Kock, M. O. de; Erwin, D. H.; Garrison, G. H.; Kirschvink, J. L.; Smith, R. Abrupt and Gradual Extinction Among Late Permian Land Vertebrates in the Karoo Basin, South Africa. *Science* **2005,** *307,* 709–714.

Wignall, P. B. Large Igneous Provinces and Mass Extinctions. *Earth-Sci. Rev.* **2001,** *53,* 1–33.

Young, G. A.; Elias, R. J. Latest Ordovician to Earliest Silurian Colonial Corals of the Eastcentral United States. *Bull. Am. Paleontol.* **1995,** *108,* 1–148.

PART II
ENERGY

"When coal came into the picture, it took about 50 or 60 years to displace timber. Then, crude oil was found, and it took 60, 70 years, and then natural gas. So it takes 100 years or more for some new breakthrough in energy to become the dominant source. Most people have difficulty coming to grips with the sheer enormity of energy consumption."

—**Rex Tillerson**

CHAPTER 5

Energy and Electricity

SANJIB KUMAR SAHOO*

Department of Electrical and Computer Engineering, National University of Singapore, Singapore 117570, Singapore

*E-mail: elesahoo@nus.edu.sg

ABSTRACT

This chapter gives an overview of electricity generation and where methods of generation of electricity from various primary energy resources are illustrated. The electromagnetic alternator driven by a turbine is the fundamental mechanism to produce electricity. Traditionally, fossil fuel-based energy resources are used to drive this turbine. As burning of fossil fuel are environmentally polluting and are proven to accelerate global warming and drive climate change, renewable and nuclear resources are increasingly replacing them. While nuclear power plants have been subject to radiation leak with catastrophic consequences, renewables like hydro, solar, and wind resources are being harvested to generate electricity in sustainable and environmentally friendly manner. Electric vehicles are the future of modern transportation and this requires an efficient storage and distribution system for electricity. With advances in information and communications technology, we are moving toward having smart grids which not only improves connectivity of the power produced by multiple energy resources but also optimizes energy production by efficiently managing the dynamic understanding of the usage patterns in a distribution network.

5.1 INTRODUCTION

Energy is defined as the capacity to do any activity in nature. Over the millennia, human beings have been improving their ability to harness energy from nature to improve their lifestyles. Early human beings burnt fuelwood for giving them warmth, protection, and cooking food. In the 19th century, steam engines were developed to convert heat energy into motion which in turn led to the industrial revolution in Europe. Steam engines heat water in a boiler to transform it into steam at a high pressure. The high-pressure steam is then allowed to expand through the piston or turbine, causing mechanical work. The steam engine produced continuous rotary motion and was used in a wide range of manufacturing machinery to be powered. The engines needed coal or wood fuel to be burnt for producing the required heat for the steam engines for locomotive and mills. In the late 1850s, internal combustion engine was invented where combustion of fuel along with air occurs inside a combustion chamber resulting in high-temperature and high-pressure gas. The force produced in the process is applied to pistons. Such engines are used in the majority of vehicles today.

In the early 1800s, Michael Faraday demonstrated the principle of electric motors. Electric motors converted electrical energy into mechanical energy. Electric motors could be used to provide power to individual machines. This eliminated the complicated belt and shaft system for power distribution and improved ease of control and overall efficiency. In the late 1800s, Thomas Alva Edison developed an electric light bulb. Before the light bulb became popular, lamps were burning oil or gas to produce light. These lamps were dangerous, releasing harmful fumes. Today most illumination is done through electric lamps of many designs.

Electricity is clean at the point of use. Electricity is a secondary form of energy. It is relatively simple to convert other forms of energy into electrical energy and vice versa. It can be produced at a place far away from human habitations and transmitted over hundreds of kilometers and distributed over a large number of points of use. It can be transported from source to point of use almost instantaneously.

There are four major sectors using energy: residential, commercial, transportation, and industrial. Residences use energy for heating/cooling, lighting, and other household work. Although natural gas is commonly used for heating or cooking, all other tasks use electrical energy. A commercial sector like offices and malls use an almost equal share of natural gas and electricity. Industrial sector uses an equal share of coal, oil, natural gas, and electricity. The transport sector mostly uses oil and some natural gas. However, of late more electrification of the transport sector is happening with the mass transport sector like railways being mostly electrified. An individual passenger vehicle is also being electrified at a large rate.

5.2 ELECTRICITY GENERATION

Electricity is generated in many different ways. The common method is known as electromagnetic-mechanical energy conversion. A coil of conducting wire is rotated between two poles of a magnet as shown in Figure 5.1. According to Faraday's law, the changing magnetic field seen by the rotating coil causes a voltage to be induced in it. This induced voltage can then supply current to an electrical load. The coil is rotated by a turbine which basically converts the mechanical energy into electrical energy. The turbine can receive the required mechanical energy from various sources. A thermal power station burns fossil fuel and converts the heat energy to provide the mechanical energy for the turbine. A nuclear generator uses the heat generated from fission to produce the heat which is then converted to mechanical energy for the turbine. A hydroelectric plant uses the potential energy of water stored in a dam to flow into the turbine. A wind generator uses the kinetic energy of air to rotate the turbine. However, there are other sources like solar photovoltaic (PV) where sun rays are directly converted to electricity by the PV cells. A fuel cell converts the fuel to electricity directly. The sources of electricity can be categorized as conventional (fossil fuel based) or renewable (hydro, solar, wind, etc.).

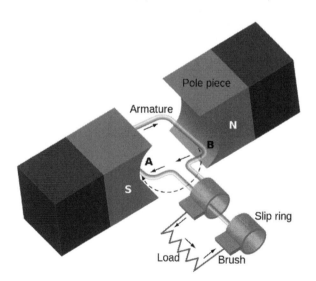

FIGURE 5.1 Basic structure of the electromagnetic generator.

Source: Wikimedia Creative Commons Library. (https://commons.wikimedia.org/wiki/File:Elementary_generator.svg)

Figure 5.2 gives a distribution of electricity generation from various sources, where the electricity generated from renewable sources is only about 20% but is expected to grow to about 30% by the year 2040. Nuclear energy accounts for about 11% of total electricity production now and is expected to grow to about 16% in the year 2040. Fossil fuels like coal, petroleum and natural gas are currently the major source accounting for close to 70% of the electricity generation but this is expected to fall to about 55% by 2040. Among the fossil fuels, coal has the major share followed by natural gas. Liquid petroleum has a very small share in the production of electricity.

5.2.1 THERMAL POWER PLANT

In thermal power plants, heat is used to transform water into steam. When steam at high pressure and high temperature is allowed to expand through the turbine, it imparts kinetic energy to it. The kinetic energy of the turbine helps in rotating the coil in the magnetic field and gets converted to electrical energy. The variations of the thermal power stations are based on the source (fuel) of heat.

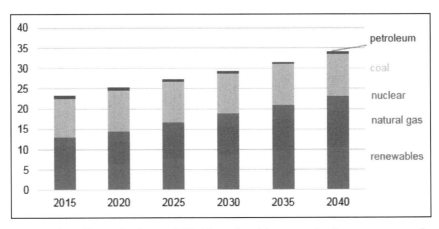

FIGURE 5.2   (See color insert.) World net electricity generation by energy source (in trillion kilowatt-hour).

Source: International Energy Outlook (2017).
(https://www.eia.gov/outlooks/ieo/pdf/exec_summ.pdf)

Figure 5.3 shows the schematic diagram of a coal-fired thermal power plant. Coal is pulverized and mixed with preheated air for combustion inside the boiler furnace. Coal contains carbon along with other impurities like sulfur, iron, and silicon. Carbon combines with oxygen in combustion to produce a large amount of heat, along with carbon monoxide and carbon dioxide as by-products. The other impurities which do not take part in combustion get collected as ash at the bottom of the furnace, although some dust particles get carried by the exhaust gases. These gases are the major pollutants of the atmosphere, with carbon dioxide contributing the most to greenhouse effect. Water is the working fluid which is heated inside the boiler as it is pumped through thin tubes exposed to the heat produced from combustion. The exhaust steam from the boiler is at high temperature and high pressure. When fed to the blades of the turbine, it expands and causes the turbine to move. The steam that exits the turbine is at less pressure and temperature. It is cooled inside a condenser so that it can be fed back to the boiler. The turbine is connected to an alternator, which is an electromagnetic generator shown in Figure 5.1 that produces the electricity.

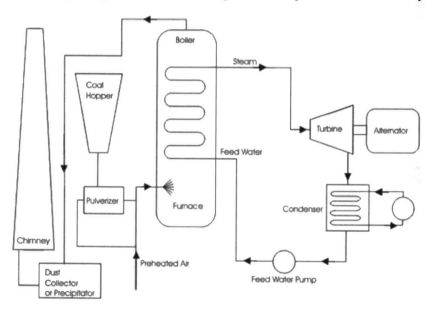

FIGURE 5.3   Schematic of a thermal power plant.

5.2.2   *FOSSIL FUELS*

Mostly fossil fuels like coal, liquid petroleum, and natural gas are used in thermal power plants to produce the heat. In the combustion process, oxygen reacts with the hydrocarbons in the fuel to produce carbon dioxide

(CO_2), water vapor (H_2O), and a large amount of heat. Coal is the most commonly used as it is most commonly available and is cheapest. However, coal produces the largest amount of greenhouse gases like carbon dioxide and contributes to pollution and global warming. Oil and natural gas are the two other types of fossil fuels used in thermal power plants. Natural gas is used in combined cycle plants where both gas turbines and steam turbines are used for improved efficiency. Natural gas produces about 30% less carbon dioxide than burning petroleum and about 45% less than burning coal, for producing the same amount of heat (Sarkar, 2015).

Fossil fuels are formed in earth's crust by decomposition of plant and animal remains over millions of years. It is considered a nonrenewable fuel as the rate of formation is far slower than the rate at which it is used. The accessible reserves for fossil fuel are projected to be finished in a few decades.

5.2.3   BIOMASS

Another fuel that can be used is biomass. Biomass is organic matter produced by natural plants through the process of photosynthesis of carbon diode from atmosphere and water from the ground, in presence of sunlight. Such biomass can be directly burned to produce heat or can be converted to biofuel before being used in combustion in power plants.

Not all heat produced from burning the fuel is converted to electricity. Part of the heat is lost to the environment as waste heat. Typical efficiency of single cycle thermal power plant falls between 33% and 48%. The efficiency of combined cycle power plants can be above 60%.

5.2.4   NUCLEAR POWER PLANT

In a nuclear plant (Murray, 2009), fuel is converted to energy without emission of pollutants. As shown in Figure 5.4, heat is generated in the reactor's core by controlled nuclear fission. The coolant which is pumped through the reactor is heated with this heat and thereby removes the energy from the reactor. The hot coolant is converted to steam for the turbine. Since nuclear fission creates radioactivity, it involves radiation hazard. To prevent the radiation from leaking into the environment, a thick concrete dome is built around the reactor. This dome also acts a protective dome against external impacts (Wagner, 2011). However, there is still some risk of core melt-down and associated radiation contamination of the surrounding environment. The used fuel rods from the nuclear reactors are still radioactive and need to be stored or disposed of safely. All these make nuclear reactors have high initial cost. Also, nuclear fuel is scarce and is available only in a few countries. Hence, the overall contribution of nuclear energy toward world electricity production is expected to grow slowly as compared to that of renewable energy.

5.2.5 HYDROELECTRIC GENERATORS

In hydroelectric plants, the potential energy of dammed water is converted to kinetic energy as it flows down a large pipe (Wagner, 2015). As shown in Figure 5.5, sluice gates at the dam can control the flow rate of water along the penstock which guides the water into the turbine. The turbine drives a generator. The power extracted from the

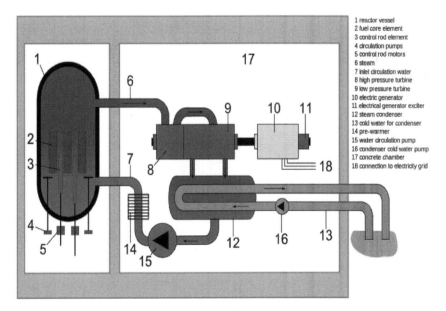

FIGURE 5.4 Schematic of a nuclear power plant.

Source: Robert Steffens. (https://commons.wikimedia.org/wiki/File:Boiling_water_reactor_english.svg)

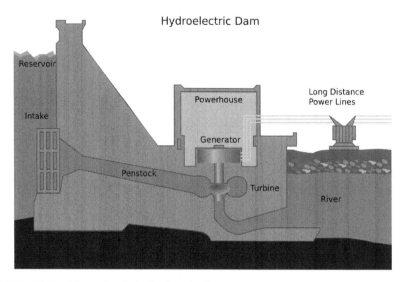

FIGURE 5.5 Schematic of a hydroelectric plant.

Source : Wikimedia Creative Commons Library. (https://commons.wikimedia.org/wiki/File:Hydroelectric_dam.png)

water depends on the flow rate and on the difference in height of the dam and location of the turbine. This height difference is called the head. Hydroelectricity is renewable and clean as it does not produce any pollutants. These plants can be made available at short notice. Also, a pair of reservoirs upstream and downstream can work together for pumped storage. During the time of peak demand, water flows from the top reservoir through the turbine to the bottom reservoir generating electricity. When there is excess power available, pumps are used to pump water from the bottom reservoir.

The typical efficiency of hydroelectric plants is close to 90% and it is a renewable energy source. However, hydropower plants require the construction of large dams which may submerge vast areas of forests, damage environment and can displace people. Initial cost for such project is also fairly high.

5.2.6 WIND GENERATOR

Wind generator works on the opposite principle as fans. While fans use electricity to produce wind, the wind generator produces electricity from the wind. Wind consists of the bulk movement of air over earth's surface. Figure 5.6 shows the components of a wind generator. The kinetic energy of moving air particles is imparted to the blades and rotates the turbine. The wind turbine rotates at a low speed. A gearbox is used to convert the low speed of the wind turbine to high speed at the shaft of the generator. This produces sufficient voltage at the output of the generator. To achieve maximum energy output, the turbine blades must be looking into the wind direction. This is achieved by using a wind vane and the yawing motor. The blade pitch is another important operating parameter in wind generators. The blade pitch is same as the angle of attack in propellers and it controls the incidence of the wind on the blades. The blade pitch is adjusted in accordance with the wind speed to keep the output power within the allowable range. When wind speed is below the minimum speed or too high, the pitch is kept zero. To account for varying wind speeds, the wind generator employs special mechanical hardware, power electronic converter, and control system to produce electricity at a constant frequency. The electricity output is then used locally or fed to the power grid. Wind velocity is more at higher altitudes as compared to at ground level. Hence a tower is erected to raise the height of the wind generator.

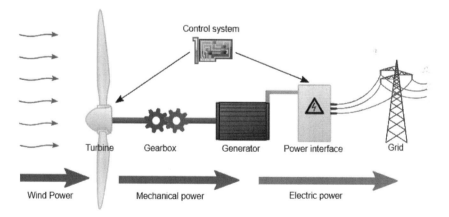

FIGURE 5.6   Wind generator system.

Source: Jalonsom. (https://commons.wikimedia.org/wiki/File:Wind_turbine_schematic.svg)

For countries having a coastline, wind farms with a large number of generators are installed offshore where the wind speed is usually steady. Wind Generator is the second highest installed capacity of renewable energy after hydroelectricity. As shown in Figure 5.7, the total installed

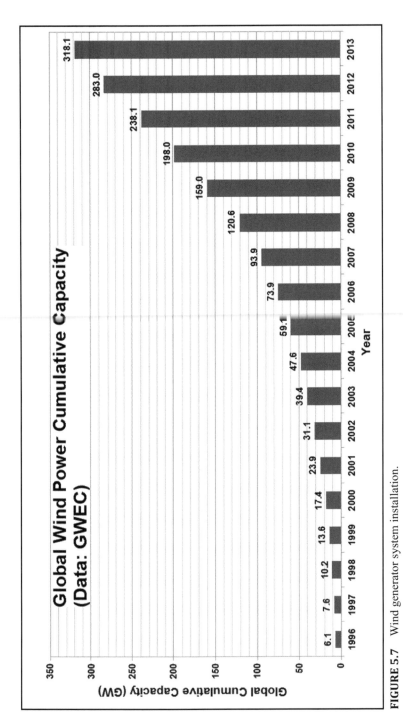

FIGURE 5.7   Wind generator system installation.

Source: S-kei. (https://commons.wikimedia.org/wiki/File:GlobalWindPowerCumulativeCapacity.png)

wind generation capacity is growing at a rapid rate and was more than 318 GW in the year 2013.

5.2.7 SOLAR PHOTOVOLTAIC

In solar PV systems, sunlight is directly converted into electricity using semiconductor devices called PV cells. Each PV panel has many such cells connected in an array. The typical conversion efficiency of solar PV panel is 20%. Hence with solar insolation of 1000 W/m^2, a solar PV panel with a square meter of the area will produce about 200 W. The solar PV panel gives a DC output. Often, a power electronic converter (inverter) is used to convert the DC into AC which can then be tied to the power grid (Fig. 5.8).

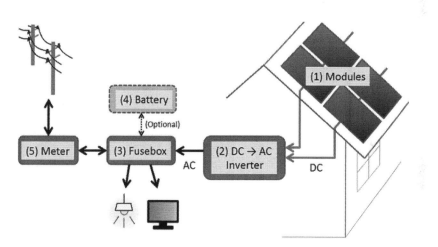

FIGURE 5.8   Grid-tied solar PV system.

Source: (https://commons.wikimedia.org/wiki/File:PV-system-schematics-residential-Eng.png)

Household users can have low power standalone solar PV systems with a charge controller and battery. Such systems are very useful when the grid is not available for isolated communities in developing countries. Solar PV systems can be easily scaled up to very large solar farms also where power is supplied to grids.

Such panels are mounted on ground or rooftops and are exposed to Sun. Usually, the amount of power produced is proportional to the solar insolation. Hence, sun-tracking systems are used to ensure that the panels face the sun all the time. Such tracking systems increase the power produced by 20–50%. Solar PV is the third among renewable energy sources after hydro and wind. As shown in Figure 5.9, the installed capacity in the world is increasing at a large pace due to drop in prices of PV panels along with improvement in technology and increase in the scale of manufacturing. In the year 2017, the projection is that cumulative solar PV installations worldwide would reach 401GW.

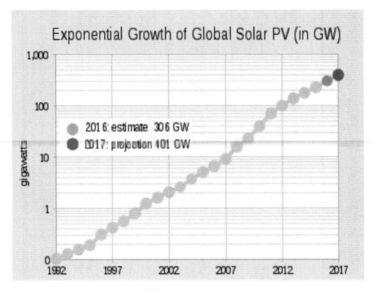

FIGURE 5.9   Global solar PV installation.

Source: Rfassbind. (https://commons.wikimedia.org/wiki/File:PV_cume_semi_log_ chart_2014_estimate.svg)

5.3   ELECTRICITY TRANSMISSION AND DISTRIBUTION

A variety of facilities can generate electricity. These electricity genera- tors can be at various geographical locations and far away from the users. These generators are of varying characteristics in terms of their avail- ability. Coal-fired and nuclear power plants usually have a constant output. The output of hydroelectric and natural gas fired plants can be adjusted quickly. Renewable energy sources such as wind and solar photovoltaics are usually intermittent. Similarly, the loads over a large geographical area have a wide variation as different sectors require energy at different times of the day. For example, the residential loads increase when people are at home. Commercial and industrial loads mostly start in the morning and are present till late evenings. Thus, it is desirable to consolidate the generators and loads with all connected together in what is known as the "electrical grid." The electrical grid consists of generators, transformers, and transmission lines, as shown in Figure 5.10.

5.3.1 THREE-PHASE POWER

Most household electric supplies have a live and neutral connection, constituting a single phase supply. The voltage and current are sinusoidal functions with the frequency of 50 or 60 Hz. However, bulk generation of electricity is done as three-phase alternating current. The three-phase power has three sinusoidal voltages with phase difference of one-third of the period. This phase delay among the phases causes the total power supplied by the generator to be constant in time. Also, the three-phase supply uses less conductor for the same amount of power rating. This leads to a large saving in the initial cost of transmission systems.

5.3.2 TRANSMISSION LINES

The power output from the generators has to be carried to the user which may be very far away. Electric power is defined as the product of voltage and current. The current flowing in a conductor results in Ohmic loss or restive loss which is the product of the square of the current and resistance of the conductor. The resistance of the conductor depends on the resistivity of the conductor material, its cross-sectional area, and length. Commonly aluminum or copper is used for such conductors. The resistance of the conductor increases proportionally with length. For transmission lines with long distance, the resistance can be substantial. Hence, the current should be reduced. This is achieved by stepping up the voltage using step-up transformers. Transmission-level voltages are typically at or above 110,000 V or 110 kV, with some transmission lines carrying voltages as high as 765 kV. The transmission lines are either overhead bare conductors or underground power cables. Overhead transmission lines are vulnerable to weather conditions. The underground cables are free from disturbances in weather conditions but are expensive to install and maintain.

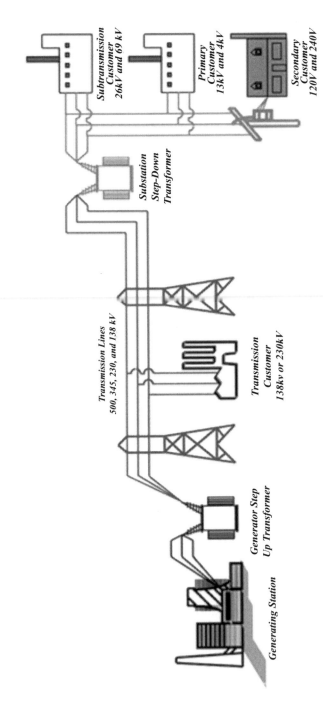

FIGURE 5.10   **(See color insert.)** Basic structure of the electric system. Blue lines are transmission lines, green are distribution, and black lines represent power generation lines.

5.3.3   DISTRIBUTION

After bulk power is transported over the transmission lines, it is distributed using the distribution network. First, the high transmission voltage is stepped-down to the medium voltage at the transmission substation. The high power industrial and commercial loads are supplied at medium voltage like 66 kV. Another stage of stepping down using the distribution transformer is done before supplying power to the residential loads. Usually, a neutral wire from the secondary side of the distribution transformer is used to supply single phase supply to each residential unit as most household appliances are single phase loads. Households are grouped together and distributed among the three phases to ensure balanced load on each phase.

5.3.4   SMART GRID

As technology changes and better options become available, significant improvements could be made to the electricity grid. For example, energy storage technologies could allow electricity to be stored for use when demand for electricity peaks or increases rapidly, increasing efficiency and reliability. Newer, more advanced meters allow better data collection for more effective management and faster response times. Even vehicles could play a role, as smart charging can allow electric cars to interface with the electric grid (Fig. 5.11).

Distributed generation systems, such as solar panels on individual homes, reduce the distance that electricity has to travel, thereby increasing efficiency and saving money. Investments made by consumers—such as purchasing energy-efficient appliances, constructing more energy-efficient buildings, or installing solar panels—save customers money and utilize energy more efficiently at the same time.

5.3.5   ENERGY STORAGE

Sometimes, it is essential to store electrical energy for later use. Renewable energy sources like sun and wind are intermittent. Sun is only available during daytime wind speed and direction to keep varying. In such cases, it is essential to generate energy when it available and store it for later use. The most common methods for large-scale energy storage are as follows:

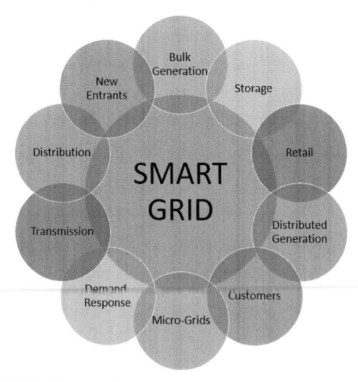

FIGURE 5.11   The smart grid.
Source: Adapted from PWC.

5.3.5.1   PUMPED STORAGE

In this method, two large reservoirs are used. When excess power is available, water is pumped to the top reservoir where energy stored as potential energy of water. Later, when power demand goes up, water from the top reservoir is allowed to drain through a turbine generating electricity. The round trip efficiency for energy storage falls between 70% and 80%. More than 90% of the bulk energy storage of electricity uses this method.

5.3.5.2   RECHARGEABLE BATTERY

The most common method of storage of a small to medium amount of electricity is done through rechargeable batteries. Portable devices cannot be connected to the electric grid. The electric battery can store energy in the form of chemical energy and can supply it when needed. Many personal gadgets and portable devices like mobile phones, laptop notebooks, tablets, etc. run on batteries. Electric vehicles use batteries to store enough energy which is then supplied to run the electric motors of the car. There are many technologies of batteries—the most common being lead–acid, lithium–ion. Lead–acid batteries are cheap but suffer from low energy per weight.

Primary (single-use or "disposable") batteries are used once and discarded; the electrode materials are irreversibly changed during discharge. Common examples are the alkaline battery used for flashlights and a multitude of portable electronic devices. Secondary (rechargeable) batteries can be discharged and recharged multiple times using an applied electric current; the original composition of the electrodes can be restored by reverse current. Examples include the lead–acid batteries used in vehicles and lithium-ion batteries used for portable electronics such as laptops and smartphones.

Batteries come in many shapes and sizes, from miniature cells used to power hearing aids and wristwatches to small, thin cells used in smartphones, to large lead–acid batteries used in cars and trucks, and at the largest extreme, huge battery banks the size of rooms that provide standby or emergency power for telephone exchanges and computer data centers.

According to a 2005 estimate, the worldwide battery industry generates US$ 48 billion in sales each year with 6% annual growth.

Batteries have much lower specific energy (energy per unit mass) than common fuels such as gasoline. This is somewhat offset by the higher efficiency of electric motors in producing mechanical work, compared to combustion engines.

5.3.5.3 ENERGY EFFICIENCY

Energy efficiency has been defined as "the first fuel," as it is the one energy resource that all countries possess in abundance. Strong energy efficiency policies are therefore vital to achieving the key energy-policy goals of reducing energy bills, addressing climate change and air pollution, improving energy security, and increasing energy access (IEA, 2016).

Globally, energy consumption and economy development have been decoupling, with the gross domestic product (GDP) increasing by more than 90% between 1990 and 2014, while total primary energy supply (TPES) grew by 56%.

5.4 COST OF ELECTRICITY

Electricity can be generated from many different sources, mainly categorized as the fossil fuels and renewable sources. Although the fossil fuels
are the major source for electricity generation at present, the global
warming and climate change due to greenhouse gas emission from the
power plants has become a major concern. The push for more and more
renewable energy sources, however, face the challenge of cost. Hence,
it essential to understand the cost of electricity generated from various
sources to understand the future of electricity generation. Cost of electricity is usually given as cost per kilowatt-hour or megawatt-hour.

A consistent measure known as levelized cost of electricity (LCOE)
is used to compare the economics of various sources. It is obtained by
adding all costs starting from the cost of building the generating plant, the
cost for operation and maintenance of the plant over its complete lifetime
and other costs associated with it; diving it by the total amount of energy
produced from it. All costs must be converted to its net present value
(NPV). The LCOE is the price at which the generated electricity must be
sold to achieve break-even. The LCOE for new investments in renewable
energy sources must be comparable with the existing cost of electricity
purchased from the grid (Fig. 5.12).

Three important parts of the cost are as follows:

1) Capital cost—share of this is low for fossil fuel based plants with
   mature technology, but is high for renewable energy sources like
   solar PV, wind power, and nuclear
2) Fuel cost—this cost is high for fossil fuels and biomass, low for
   nuclear but zero for renewable sources
3) Operation and maintenance cost to run the plant equipment, and
   cost of energy to run the auxiliary machinery

5.5 FUTURE OF ELECTRICITY

The demand for electricity is rising at an ever-increasing rate for a variety
of reasons. As the population grows and the world gets more industrialized
and urbanized, demand for energy, in general, will grow. Due to the ease
of use, electricity is the most common form of energy used. Although most
electricity is generated by combustion of fossil fuels, there is a growing

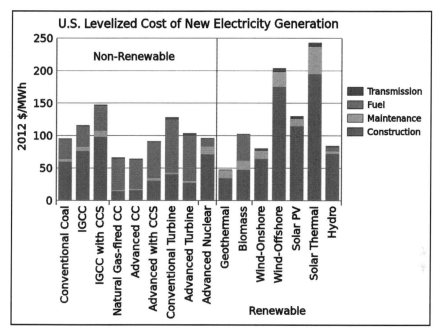

FIGURE 5.12 (See color insert.) Estimated levelized cost of electricity generation (in $/megawatt-hour) in 2018 based on 2011 statistics.

Source: Delphi234. (https://commons.wikimedia.org/wiki/File:LCOE_US_Plants_Coming_Online_in_2019.svg)

concern for global warming due to the emission of greenhouse gases like CO_2. Electricity generated from green and renewable sources holds the promise for future energy.

The electricity landscape is undergoing a transformation, becoming more complex than ever before with improving technologies, falling costs of renewable energy sources reaching grid-parity, and changing regulatory policies in electricity markets. There is increasing electrification, that is, more and more users are being converted into electricity. For example, the conventional natural gas based space heating is converted to electrical heating. The electricity generation is being decentralized, with smaller generators being added closer to the load instead of huge central power plants. Finally, with faster communication bandwidth, the grid itself is being digitized with grid protection, switching and control being done digitally. The smart and connected technologies are invading conventional electric power grid. Distributed generation, as well as storage, smart meters, and increasing adoption of electric vehicles, are impacting the electricity system.

Due to the increasing availability of renewable energy sources, electrification will reduce our reliance on fossil fuels. In many cases, electrification will also increase energy efficiency.

The most promising electrification opportunities are in those segments that are among the largest polluters: transportation, commercial applications, and residential heating, and cooling. Decentralization takes the power supply and storage away from the main grid and into locations closer to where it is needed. There are various advantages to this, such as reducing losses of energy during transmission and lowering carbon emissions. Blackouts will be reduced as the security of supply is increased, thanks to the larger number of available power sources. Decentralization also enables control of energy use during peak-demand and high-pricing periods.

KEYWORDS

- electricity
- thermal power plant
- hydroelectric generator
- renewable energy
- solar PV
- smart grid
- electrification of transportation

REFERENCES

Gao, D. W. *Energy Storage for Sustainable Microgrid*; Elsevier: Amsterdam, 2015.
Mihail, H. *Technologies for Electrical Power Conversion, Efficiency, and Distribution*; Engineering Science Reference: Methods and Processes: Hershey, PA, 2010.
Momoh, J. *Smart Grid: Fundamentals of Design and Analysis*; John Wiley & Sons, 2012.
Murray, R. L. *Nuclear Energy: An Introduction to the Concepts, Systems, and Applications of Nuclear*, 6th ed.; Elsevier: Boston, 2009.
Sarkar, D. K. *Thermal Power Plant: Design and Operation*; Elsevier: Amsterdam, 2015.
Tooraj, J.; Pollitt, M. G. *The Future of Electricity Deman: Customers, Citizens, and Loads*; Cambridge University Press, Cambridge, 2011.
Wagner, H. *Introduction to Hydro Energy Systems: Basics, Technology, and Operation*; Springer-Verlag: Berlin, 2011.

CHAPTER 6

Unconventional Fossil Fuels

DEEPAK DASH[1,*], LOKANATH PEDDINTI[1], and
SANDEEP NARAYAN KUNDU[2]

[1]*Reliance Industries Limited, Thane-Belapur Road,
Navi Mumbai 400701, Maharashtra, India*

[2]*Department of Civil & Environmental Engineering,
National University of Singapore, Singapore*

Corresponding author. E-mail: drdash97@gmail.com

ABSTRACT

Fossil fuels play a dominant role in global energy system. With an ease to develop and extract conventional resources fast depleting, unconventional fossil fuel resources shall play an important role in the future. With advances in technology, like horizontal drilling and hydrofracturing, unconventional resources like shale gas and coal seam gas are increasingly playing a bigger role in the present-day footprint of fossil fuels. It is therefore imperative to understand their settings, occurrence, and global distribution of these unconventional energy resources. The present chapter is an attempt to present an overview of the various fossil field resources available on earth, their geographical distribution and resource potential, their potential prospects, and technological challenges facing their extraction and exploitation.

6.1 INTRODUCTION

Energy demand is increasing worldwide, specifically driven by strong economic growth and development in populous developing nations. Fossil fuels are nonrenewable resources, and according to the World Energy

Council comprise about 87% of all our energy demands today. Renewables account for the remaining 13%, of which nuclear and hydropower comprise 12%. Solar, wind, and geothermal energy contributes to barely 1% in today's energy mix. Fossil fuels are classified as conventional and unconventional depending on their geological occurrence and mode of extraction. Conventional fossil fuels occur in "conventional" reservoirs which can be discovered and developed with vertical and directional wells. These reservoirs typically have a hydrocarbon accumulation with a distinct oil–water or gas–water contact. Hydrocarbons are also trapped in formations where they cannot be extracted through conventional means. Recent developments in technology have advanced fossil fuel extraction from such formations called unconventional reservoirs where its huge commercial potential is being explored and exploited globally. Collectively known as unconventional fossil fuels (Fig. 6.1), these forms of energy are now replacing the already depleting conventional fossil fuel resources. In this chapter, we shall learn about the fundamental aspects of unconventional fossil fuels which involve the types, geological occurrence, and global distribution of unconventional energy and understand their future resource potential.

6.2   DEFINING UNCONVENTIONAL FOSSIL FUELS

The jointly issued statement by American Association of Petroleum Geologists (AAPG), the Society of Petroleum Engineers (SPE), World Petroleum Congress (WPC), and Society of Petroleum Evaluation Engineers (SPEE), in 2007, defines "unconventional resources" as "petroleum accumulations pervasive throughout a large area and not significantly affected by hydrodynamic effects." These types of accumulations are also referred as "continuous-type deposits." According to Canadian Association of Petroleum Producers (CAPP), conventional resources are at the top of the resource triangle and are easy to develop but increasingly harder to find owing to their localized and limited occurrence. Unconventional resources, on the other hand, are found in large volumes but are more difficult to develop.

Unconventional hydrocarbon resources are distinct from conventional hydrocarbon resources. The unconventional resources are not defined by the elements of the petroleum system (source, reservoir, trap, seal, migration, and timing). For example, shale gas and shale oil are produced from a source rock, with no need for a reservoir, trap, or seal. Oil occurs in the

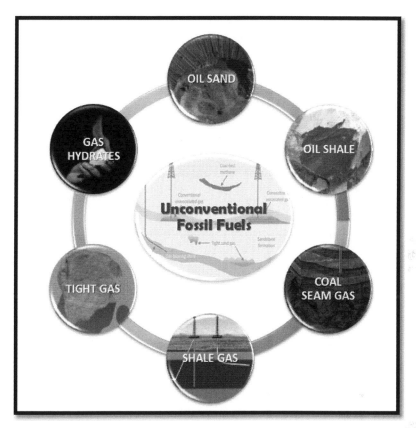

FIGURE 6.1 Unconventional fossil fuels and its types.

solid state in oil sands, so the presence of a trap or seal is not needed for its accumulation.

The term unconventional fossil fuel or unconventional hydrocarbon is generally used to describe fuels that cannot be extracted using conventional drilling or mining. They often involve new advanced extraction technologies such as horizontal drilling, directional drilling, and hydraulic fracturing (fracking) for extraction. Oil and gas reservoirs in which wells can be drilled so that it can flow naturally or can be pumped to the surface are commonly referred to as "conventional" oil and natural gas. The hydrocarbon in unconventional reservoirs does not flow naturally to the surface. This condition occurs because shale is so fine grained (the pores are so small) that surface tension, capillary forces, and the presence of mixed wettability (of the minerals and organics) in the reservoir yield an extremely diffuse distribution of fluids (water, oil, and gas). Both unconventional and conventional fossil fuels differ in their geologic occurrences and accessibility; unconventional fuels are found within pore spaces in a wide variety of geological formations, whereas conventional fuels are often found localized in high concentration in geological reservoirs which are easy to access and extract. While hydraulic fracturing of the shale helps to liberate the hydrocarbons from the groundmass of the shale, the length of the horizontal well makes the whole operation economically feasible. A vertical well, drilled into the shale, would not be economically feasible.

6.2.1 TYPES

Unconventional resources typically fall into five main categories on the basis of hydrocarbon accumulations. These accumulations of solid, liquid, or gaseous hydrocarbons have been generated over geological time from organic matter in source rocks. The character of these hydrocarbons depends on the depositional environments and geological settings. The main categories are shale oil, shale gas, tight sand gas and oil, coalbed methane (CBM), and oil sands (Fig. 6.2).

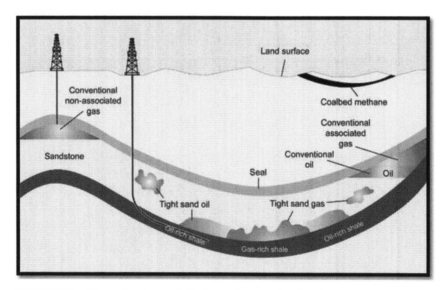

FIGURE 6.2 **(See color insert.)** Geology of conventional and unconventional oil and gas.

Source: US Energy Information Administration. (https://www.eia.gov/todayinenergy/detail.php?id=110)

6.2.2   ASSOCIATION WITH SHALE

Most unconventional fossil fuels occur in the rock shale. Shale is geologically defined as any "laminated, indurated (consolidated) rock with >67% of clay-sized materials" (Jackson, 1997). Shale comprises almost 50% of all sedimentary rocks found on earth. Shale is described as extremely fine-grained particles, typically less than 4 μm in diameter but may contain variable amounts of silt-sized particles (up to 62.5 μm). No two shales are alike and they vary aerially and vertically within a trend and even along horizontal wellbores (King, 2010). Shales are fine-grained clastic rocks which are deposited in low-energy sedimentary depositional settings like lakes and far offshore. When organic matter is present in high concentration, they tend to have high organic carbon which can produce hydrocarbons on maturation. Shales, therefore, are hydrocarbon source rocks from which most conventional oil and gas originates. A shale resource system is described as a continuous organic-rich source rock that may be both a source and a reservoir rock for the generation and production of petroleum (oil and gas) or may charge and seal petroleum in juxtaposed, continuous organic-lean intervals. There are two basic types of producible shale resource systems: gas- and oil-producing systems with overlap in the amount of gas versus oil. Dry gas resource systems produce almost exclusively methane; wet gas systems produce some liquid hydrocarbon, and oil systems produce some gas. These are commonly described as either shale gas or shale oil, depending on which product predominates in production. The industry commonly describes these as shale plays or unconventional hydrocarbon plays. Shales act as the source, the reservoir, and also as the seal in unconventional hydrocarbon plays. The characteristics of this unconventional fuel are as follows:

- good thickness;
- high organic richness (3–5%) (Mishra et al., 2012);
- higher maturity value (0.9–1.3 Ro);
- good gas saturation and greater transformation ratio (TR, >90%);
- abundant gas storage capacity (40–200 BCF/section);
- low matrix permeability less than 0.001 milliDarcy (mD)'
- requisite brittleness for hydrofracturing;
- regional in extent with diffuse boundaries, low matrix permeability, and low recovery factors;
- development is highly cost-intensive, however, with a very long well life predicted to exceed 25–30 years, compared to 8–10 years for traditional gas wells;

- there is a need for a much higher number of wells, state of the art technology (e.g., horizontal drilling and modern fracturing techniques), and large developments; and
- high pore pressure.

6.3   SHALE OIL

Shale oil is defined as an unconventional oil found in sedimentary rock characterized by very low permeability—typically shale. There is also evidence of its occurrence in organic-rich mudstones that have generated, migrated, and stored oil into juxtaposed and continuous intervals. This also includes interbedded, underlying, or overlying sediments which may contain low organic matter (EIA, 2011). A number of formations may contain very little shale lithology and/or mineralogy but are considered to be shale by grain size only. An organic-rich source rock which is buried deep in the oil maturation window is most likely to produce oil. The oil maturation window, or simply the oil window, is the depth range at which adequate temperature conditions are reached for maturation of organic matter in sediments to form oil. On maturation, oil is generated and expelled from the source rocks to the formation. The temperature ranges between 60°C and 120°C (approximately 2–4 km depth) for oil window, while the temperature ranges between 100°C and above 200°C (3–6 km depth) for the gas window. Some of the oil and gas still remains in the source rock which is most difficult to produce unless it was open. Shale oil has the following geological characteristics:

I.   Shale is mainly deposited in the lacustrine depositional environment, with Type I and Type II kerogen. The properties of matured shale source rocks are as follows:

- vitrinite reflectance (Ro) ranges from 0.7 to 1.1%;
- total organic carbon (TOC) ranges from 1.4 to 25.6%;
- free hydrocarbons (S1) ranges from 1.2 to 11.6 mg/g; and
- pyrite content 5.4–34.5%.

Kerogen is the solid insoluble component of the organic material. Type I is the hydrogen-rich organic material and Type II is the marine organic-rich material and both are oil prone. Ro is the thermal maturity indicator. TOC is the total amount of organic carbon present in a rock which is the determining factor in a rock's ability to generate hydrocarbons. S1 is the free hydrocarbons present in the rock sample before the analysis. The hydrocarbon generation potential of shale is five times more than that of mudstone.

II. Shale has a laminated structure where parallel bedding planes impart fissility to the rock which also acts as the space to contain hydrocarbons.

III. Shale has higher brittle mineral content like quartz and feldspar (~40%) with clay minerals lower than 50%.

IV. In shales, the residual liquid hydrocarbon occurs as adsorbed and in the free form. Adsorption occurs in organic matters, and free gas occurs confined in nanosized pore throats in pyrite, clay and brittle minerals, and parallel bedding fractures.

On the basis of origin, three main types of shale occur on earth. These are marine shale, lacustrine shale, and terrestrial shale. Terrestrial shales have organic origins similar to coal-forming swamps; lacustrine shales have origins in fresh or brackish water; and marine shales have origins in salt water life.

6.3.1 MAJOR SHALE OIL PRODUCERS

Deposits of organic-rich shale are found in many areas around the world. US Energy Information Administration (EIA) identified 42 countries representing 10% of the world's technically recoverable crude oil resources. They have also identified 137 shale formations in these countries. In North America, the largest oil shale deposits are in Colorado, Wyoming, and Prudhoe Bay Utah. Also, large areas of the Russia, Argentina, and China are known to have shale oil reserves. There is some prominent shale oil plays in the USA, of which the Bakken Shale Play and Eagle Ford Shale Oil Play are significant.

The Bakken Shale Play extends from the Eastern Montana and Western North Dakota to parts of Saskatchewan and Manitoba in the Williston Basin (Fig. 6.3). Although oil was initially discovered in this area in 1951, it is only recently that unconventional resources are being commercially produced owing to the development of horizontal wells and fracking technology. The Bakken shales were deposited during late Devonian and early Mississippian period. Stratigraphically, this formation consists of an upper shale layer, middle dolomitic layer, and a lower layer of shale. The shale layers are hydrocarbon source rocks as well as seals for the layer known as the Three Forks (dolomite) or Sanish (sands) Formations.

The Eagle Ford Shale Formation is located in South Texas and part of Mexico and has been a prolific world-class oil and gas play (Fig. 6.4). Cretaceous in age, this play ranges from 15- to 120-m thick shales which are stratigraphically below the Austin Chalk and above the Buda Limestone formation. At depocenters, the thickness of this shale can be over 300 m. The shale steeply dips into the subsurface toward the Gulf of Mexico and reaches to depths of over 4260 m below the sea level. The occurrence of oil and natural gas within the Eagle Ford Shale is only at depths beyond 1225 m, where the shale has been exposed to sufficient heat and pressure to convert some of the organic material into oil. At greater depths, natural gas is formed, whereas at depths beyond 4250 m, the hydrocarbons have been destroyed.

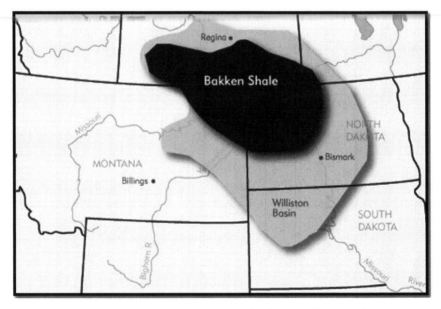

FIGURE 6.3   Distribution of Bakken shale formation in the Williston Basin.
Source: USGS.

The Permian Basin of western Texas and eastern New Mexico has become the hottest region in the US shale oil industry in 2016, as oil explorers found the largest deposit of shale oil in the Permian Basin. They have also estimated that the Wolf camp formation has the potential to hold up to 20 billion barrels of oil making it a find larger by three times to that of the Bakken formation, the largest find of unconventional oil ever discovered (US Geological Survey's Energy Resources, 2016). This gives credence to speculations that the Permian Basin could hold up to 75 billion barrels of shale oil which would make it the second largest in the world, only behind Saudi Arabia's Ghawar field.

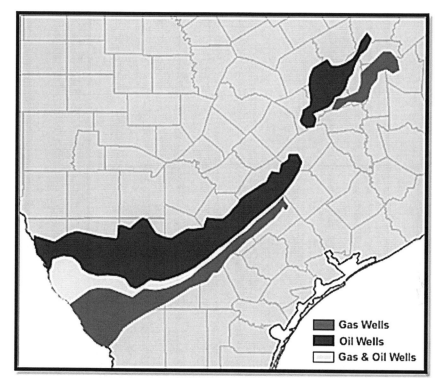

FIGURE 6.4   Distribution of Eagle Ford Shale.

Source: EIA. (https://www.eia.gov/analysis/studies/usshalegas/pdf/usshaleplays.pdf)

6.4   SHALE GAS

Shale gas is defined as a natural gas present in shale formations. It consists primarily of methane, other gases such as ethane, propane, and butane and also contains carbon dioxide, nitrogen, and hydrogen sulfide which occur in fractures and pore spaces between individual mineral grains or adsorbed onto (i.e., adhered to the surface of) minerals or organic matter within shale formation. Unlike in sandstones, pores within shales are tiny and unconnected which make it less permeable. This makes the trapped gas in shale difficult to extract.

Gas exists in organic-rich shales in three different forms (Fig. 6.5). Free gas is trapped in the pore spaces of the fine-grained sediments and adsorbed gas on the surface of organic matter and also as dissolved gas in oil and water (Curtis, 2002). The gas present in the fractures can be easily released, whereas the adsorbed gas on the surface of the organic material is the one which takes time to extract. Understanding the relative proportions of gas stored in these different forms is critical to an accurate assessment of shale gas resources (Zhang et al., 2012). The kerogen in shale is predominantly a mixture of Type II and Type III that is dominant with low hydrogen index (HI) and high TR index. Here, HI is used to compare relative maturities of a source rock, that is, shale and TR is a better proxy for relative maturity, since different source rocks may have different present day HIs.

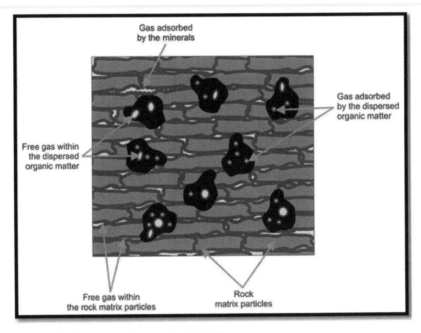

FIGURE 6.5   Gas storage in organic-rich shales.

Since ideal combinations seldom occur in nature, yet inadequacy of any one parameter may be compensated by a profusion of others in some instances. The key issue in the shale plays is the measurement of gas content and storage capacity for gas resources and reserves (Shtepani et al., 2010) and its production which is governed critically by TOC, mineralogical composition, micro fabrics, maturity, heterogeneity, in situ stress anisotropy, etc. The prolific gas production from some of the shales in the United States possess in nanoDarcy (nD) permeability systems has led to undertake researches on micro understanding of gas flow mechanism, performance, and behavior of shale reservoirs. Gas flows through a network of pores with different diameters ranging from nanometers (nm $= 10^{-9}$ m) to micrometers (μm $= 10^{-6}$ m) (Javadpour et al., 2007). The clay type, content, maturity, fracture frequency, orientations, and connectivity are other issues in placing the hydrofractures and optimizing production of shale gas. The efficacy of the gas production heavily relies on the capabilities of creating artificial fractures to enhance the system permeability by way of hydrofracturing the shale (Fig. 6.6).

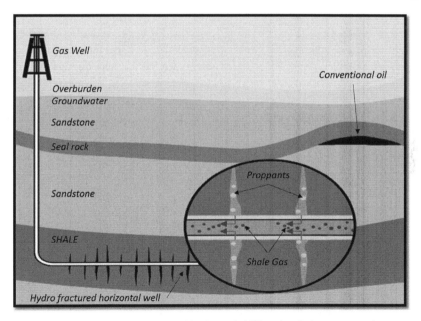

FIGURE 6.6 Hydraulic fracturing and horizontal drilling for shale gas extraction.

6.4.1 MAJOR SHALE GAS PRODUCERS

There are many countries having shale gas and oil potential. As of 2016, only commercial quantities have been established only from the United

States, Canada, China, and Argentina, and significant shale gas production is going as on date only from the United States and Canada. In the last decade, the production has come mainly from the Eagle Ford and Barnett Shale Play in Texas, Marcellus Shale in Appalachian Basin in the eastern USA and shale plays of Rocky Mountains. Experience and information gained from developing the Eagle Ford, Barnett Shale has improved the efficiency of shale gas development in the United States.

6.4.1.1 MARCELLUS SHALE

Marcellus Shale is in the Appalachian basin; it is one of the important shale plays in the eastern USA. This basin has been an important shale gas-producing province since the early 1800s. The organic-rich Marcellus Shale is recognized as a major source rock (Fig. 6.7a). Most of the studies show a general eastward thickening of the Marcellus Shale (Fig. 6.7b). The studies carried out on the organic richness or TOC content on the Marcellus Shale indicate about its ranges from less than 1% to more than 13% (wt.%), establishing the Marcellus Shale as a world-class source rock. This TOC content can also be directly related to porosity development resulting from the conversion of kerogen to hydrocarbons. Marcellus Shale, in general, had higher TOC contents in the thermogenic areas of the Appalachian Basin as compared with the other organic-rich Devonian shales.

The Middle Devonian Marcellus Shale Formation is located in the lower part of the Hamilton Group, which is bounded above by the Middle Devonian Tully Limestone and below by the Lower Devonian Onondaga Limestone (Onesquethaw Group). The Marcellus stratigraphy includes two members, which are the lower Marcellus/Union Springs Shale and the upper Marcellus/Oatka Creek Shale. They are separated by the fine-grained limestone of the Cherry Valley Member. The Cherry Valley and the Purcell limestones interpreted to be age equivalent (Lash, 2008).

6.4.1.2 BARNETT SHALE

The Barnett Shale gas play is upper Mississippian aged organic-rich shale. It is also the primary source for hydrocarbon in the Fort Worth Basin and has been the source rock of oil and gas for other conventional reservoir systems in the basin (Jarvie, 2007). The Barnet Shale extends over basin area of 14,000 km^2 and occurs at depths between 1980 and 2590 m, thickness of which ranges from 30 to 20 m. The initial production figures were

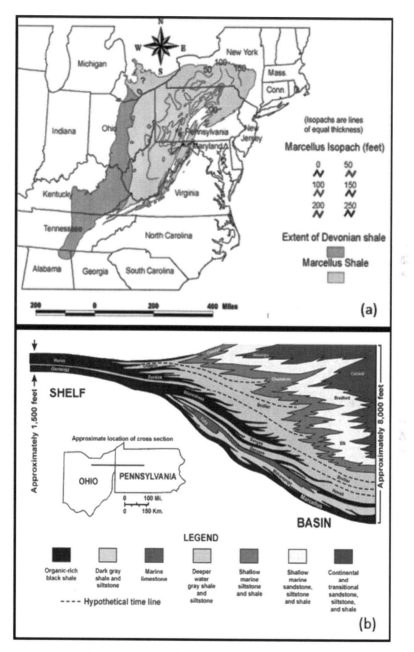

FIGURE 6.7 (a) Geographical distribution of the Middle Devonian Marcellus Shale formation. (b) West–east geological cross-section of the region.

Source: Schmid and Markowski, 2017. Used with permission.

four Million Cubic Feet per Day (MMCFD) with lateral wells extending up to 1500 m.

6.4.1.3   EAGLE FORD SHALE

The Eagle Ford Shale, in the state of Texas, was deposited during the Cenomanian and Turonian stages of the Late Cretaceous period. It is predominantly composed of organic matter-rich fossiliferous marine shales and marls with interbedded limestones making the carbonate content of the formation to be as high as 70%. The play is shallow with its formation getting shalier toward the northwestern parts (Fig. 6.8).

The Eagle Ford stratigraphy includes two depositional units. The upper and lower Eagle Ford Shale and is bounded by the Austin Chalk above and below by the Buda Formation. Eagle Ford Shale's carbonate content can be as high as 70%. High carbonate content and low clay content make the Eagle Ford more brittle and easier to stimulate for hydrocarbon production. It occurs at depths between 1220 and 3660 m and has a thickness ranging between 30 and 150 m. The shale basin extends about 10,000 km² and the formation dips toward the Gulf of Mexico (David Waldo, 2012). Its organic carbon content is between 3% and 5% with Ro above 1.0 which is well within the oil window. The shale exhibits a low porosity under 12% and permeability ranges in nD. Gas production in Eagle Ford Shale started in 2008 and the unconventional reservoir is estimated to have a condensate ratio on 50 barrels per metric million cubic feet at a pressure gradient between 0.4 and 0.7 psi/ft.

Oil explorers in the US state of Texas have discovered the largest deposit of shale oil in the Permian Basin recently triggering a rush for global giants to enter into the shale gas business in this region. According to US Geological Survey, the Wolf camp shale formation in the Permian Basin could be host to over 20 billion barrels of oil.

6.4.1.3.1   Shale Oil Resources

EIA/Advanced Resources International (ARI) carried out studies on shale oil and gas resources. According to their assessment, the total risked shale oil in-place of 5799 billion barrels, with 286.9 billion barrels, without considering US resources, as the risked, technically recoverable shale oil resource. Adding the US resource increases the assessed shale oil in-place and technically recoverable shale oil resources of the world to around 6753 billion barrels and 335 billion barrels, respectively (Table 6.1).

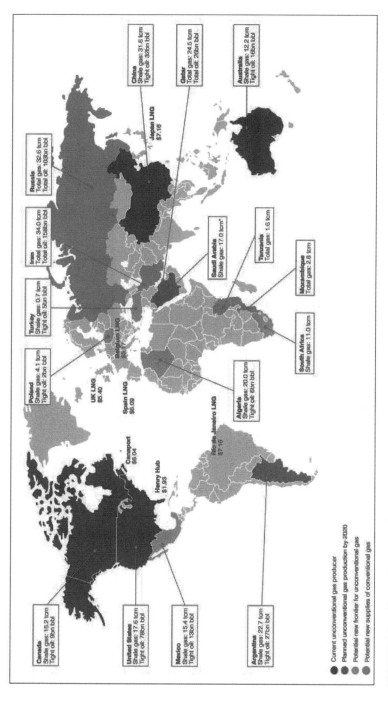

FIGURE 6.8 **(See color insert.)** Technically recoverable reserves worldwide.

Source: Reprinted with permission from World Energy Council, 2016.

TABLE 6.1 EIA/ARI World Shale Gas and Shale Oil Resource Assessment.

Region	No. of shale basins	Shale formations
Africa	**18**	**27**
Algeria	7	11
Egypt	4	4
Libya	3	5
Morocco	2	2
South Africa	1	3
Tunisia	1	2
Asia	**24**	**38**
China	7	18
India/Pakistan	5	6
Indonesia	5	7
Jordan	2	2
Mongolia	2	2
Thailand	1	1
Turkey	2	2
Australia	**6**	**11**
Australia	6	11
Eastern Europe	**9**	**11**
Other Eastern Europe	3	4
Poland	5	5
Russia	1	2
North America	**17**	**21**
Canada	12	13
Mexico	5	8
South America	**13**	**16**
Argentina	4	6
Brazil	3	3
Northern South America	3	3
Other South America	3	4
Western Europe	**8**	**13**
Other Western Europe	5	10
Spain	1	1
UK	2	2
Grand total	**95**	**137**

As shown in Figure 6.9, six countries (USA, China, Argentina, Algeria, Canada, and Mexico) host more than two-thirds of the technically recoverable shale gas resource. Only 10 countries account for over 80% of the technically recoverable shale gas resources of the world. As for shale oil, two-thirds of the technically recoverable resource is found in six countries which include the Russia, USA, China, Argentina, Libya, and Venezuela (Table 6.2).

TABLE 6.2   Risked Shale Gas in-place and Recoverable Volumes.

Continent	Risked gas in-place (Tcf)	Risked technically recoverable (Tcf)
North America (Ex. US)	4647	1118
Australia	2046	437
South America	6390	1431
Europe	4895	883
Africa	6664	1361
Asia	6495	1403
Sub-Total	31,138	6634
USA	4644	1161
Total	35,782	7795

Source: Kuuskraa, 2013.

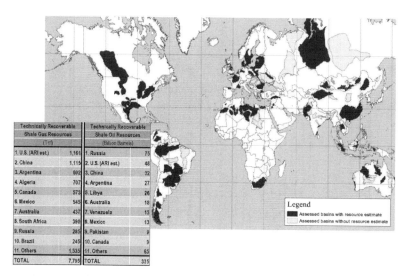

FIGURE 6.9   Assessed world shale gas and shale oil resources.

Source: Kuuskraa, 2013.

6.5   TIGHT GAS SAND

Tight gas reservoirs are defined as low-permeability reservoirs having a permeability less than 0.1 mD (Law et al., 2002). The German Society for Petroleum and Coal Science and Technology (DGMK) introduced the definition for tight gas reservoirs which includes reservoirs with average effective gas permeability less than 0.6 mD. These types of reservoirs are very difficult to produce and more energy is required to drive the fluid to the wellbore. Hence, the ultimate recovery is very low.

6.5.1   GEOLOGY

The geological features which influence the formation to be tight (impermeable but porous) are as follows. First, depositional settings like a deep basin or river bank levees are likely to host very fine sand to silt and clay facies which under lithification can form tight reservoirs. Second, the postdepositional diagenetic processes reduce the effective porosity making the rock less permeable and tight (Dutton, 1993). Third, the quartz cementation influences fracture distributions and clustering in the formation. Due to extensive cementation by authigenic clays, the matrix permeability of these sandstones is extremely low, on the order of mD (Naik, 2005).

The characteristics of tight gas reservoirs are as below:

- in situ permeability to gas of less than 0.10 mD;
- occurs in both sandstones and carbonates;
- significant formation thickness;
- isolated reservoirs inside the same formation;
- independent free water level (FWL) for every reservoir;
- overpressurized reservoir;
- water saturation below expected on usual capillary pressure curves interpretation; and
- coexistence or intercalation of source rock with reservoir rock.

Tight gas sand (TGS) reservoirs behave differently from the conventional gas reservoirs in many critical ways. Little or no water is produced from these, unlike conventional reservoirs. Tight gas reservoirs are under anomalous pressures with the gas layers varying stratigraphically from nearby fields. These reservoirs do not need a distinct top seal and exhibit an indistinct upward transition from commercial gas to non-commercial gas. The permeability of unconventional reservoir is significantly lower than the conventional reservoir (Fig. 6.10). Many extremely tight gas reservoirs may have in situ permeability as low as 0.001 mD.

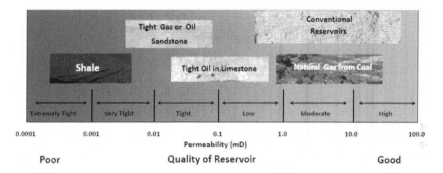

FIGURE 6.10 Permeability of conventional and unconventional reservoirs.

There is a huge potential for production of TGS. But due to micro-Darcy permeability and low porosity range, economical production of TGS is challenging. So gas production rates can be enhanced by artificial stimulation process such as horizontal drilling in combination with massive hydraulic fracturing frequently carried out to develop tight gas formations. Gas flow rates of 2–3 times those of conventional vertical wells can be achieved with these stimulation methods. Although worldwide tight gas reservoirs are known to exist in many regions, no systematic evaluation has been carried out except the United States (Fig. 6.11).

6.5.2   GLOBAL DISTRIBUTION AND RESOURCE ESTIMATES

Worldwide, TGS is believed to occur significantly only in few countries like USA, Canada, China, and Australia. At present, about 50% of daily US gas production is recovered from tight and unconventional resources. The following paragraphs show some examples of tight gas reservoirs worldwide, in which some are actively producing fields and a few are in development stage. There are many tight gas provinces in the United States and Mexico.

FIGURE 6.11 (See color insert.) The world map representing unconventional gas reserves in various regions.

Source: Pacwest Consulting.

- The Appalachian Basin (Lower Silurian);
- Medina Group (Cretaceous and Tertiary);
- Devonian age reservoirs in the East and Midwest;
- Alberta Basin;
- Michigan Basin;
- Anadarko Basin;
- Agua Nueva and San Felipe limestone (Cretaceous); and
- Burgos Basin (Eocene tight sands).

Although TGS production is growing in more than 35 countries, gas production from a TGS well is low compared to gas production from conventional reservoirs on a per-well basis. Worldwide unconventional tight gas resources (in Tcf) are shown in Table 6.3. The capital cost of unconventional gas production is high because of the need for more rigs, equipment, and people. The driving forces to bring much of unconventional TGS to market are (1) increased oil and gas prices; (2) decline in conventional oil and gas production; and (3) improvement in drilling, completion, and hydraulic fracturing.

TABLE 6.3 Worldwide Tight Gas and Coal Seam Gas Resources (Tcf).

Region	Tight sand gas (Tcf)	Coal seam gas (Tcf)
North America	1371	3017
Former Soviet Union	901	3957
Centrally Planned Asia and China	353	1215
Pacific Organisation for Economic Co-operation and Development	705	470
Latin America	1293	39
Middle East and North Africa	823	0
Sub-Saharan Africa	784	39
Western Europe	353	157
Other Asia Pacific	549	0
Central and Eastern Europe	78	118
South Asia	196	39
Total	7406	9051

Source: US National Petroleum Council, 2007.

The geological features of the tight gas fields in China have low porosity (5.2%), low permeability (0.7 mD) with good production of gas and include Changbei gas field in North Shaanxi Province and Sulige and Chuangzhong gas field. Other tight gas fields around the world are the Yucal Pacer field in Venezuela, Aguada Pichana in Argentina, Timimoun in Algeria, Aloumbe in Gabon, Sui area in Balochistan of Pakistan, and Warro field in Western Australia. The geological features of the tight gas fields in Mexico are coarse grained and compact and have low porosity but have joints and fractures to give permeability enough for economical production.

6.6 COALBED METHANE

CBM is defined as an unconventional form of natural gas found in coal seams. It mainly contains methane and along with other gasses such as CO_2 and N_2. As the name implies, CBM gas is trapped in the internal surfaces of coal structure. The gas is trapped in the coal seam in part by water pressure and in part by weak covalent bonding forces known as Van der Waals forces. This is formed during the process of coalification, which is the transformation of plant material into coal. CBM can be recovered from underground coal during, before, or after mining operations. It can also be extracted from "un-minable" coal seams that are relatively deep, thin, or of poor or inconsistent quality. Vertical and horizontal wells are used to develop CBM resources. It is accounted as one of the main resources among unconventional fuels.

CBM extraction requires drilling wells into the coal seams and removing water contained in the seam to reduce hydrostatic pressure and release absorbed (and free) gas out of the coal. A CBM well begins as a water-producing well, where the removal of the water pressure from the coal seam changes the hydrostatic pressure exerted on the methane molecules and allows them to rise naturally to the surface where they are collected. In some cases, hydraulic fracturing of the coal seam is carried out in order to expedite the process of CBM liberation. The amount of methane gas which could be kept in a cubic foot of CBM reservoirs is about six or seven times of gas volume in a cubic foot of conventional gas deposits.

Gas in coal is stored in many ways: as a free gas within the fractures and cracks of the coal bed; as adsorbed gas molecules on the internal, at the surface micropores and fractures within the coal bed; and as dissolved in solution (water) within the coal bed. The characteristics of CBM are large aerial extent with gas occurring in adsorbed and absorbed forms in coal seams. Its recovery starts with huge water production before the formation can be depressurized for the adsorbed gas to be released, and therefore the completion technique for the gas wells are critical. The huge amounts of coal within the coal-bearing basins of the world have generated significant amounts of CBM.

6.6.1   MAJOR PRODUCING REGIONS

All those countries which are producing coal are also contributing to CBM production significantly. The countries are USA, Australia, Canada, UK, France, Germany, Poland, Czech Republic, Ukraine, Russia, and China. Some of the prominent CBM basins of the world are shown in the Figure 6.12. Out of these major basins, two prominent basins, such as San Juan Basin and Powder River Basin, are described.

6.6.1.1   SAN JUAN BASIN

The San Juan Basin is located in the east-central part of the Colorado Plateau in northwestern New Mexico and southwestern Colorado (Fig. 6.12). It is one of the leading producers of CBM in the world. The upper cretaceous

Fruitland formation is the main producer of CBM in the San Juan Basin. It produces more than 2.5 Bscf/D from this formation, which is estimated to contain 43–49 Tscf of CBM in place (Ayers, 2003). Geologically, the Fruitland coals are the thickest and occur in northwest/southeast trending belt located in the northeastern part of the basin. Total coal thickness in this belt locally exceeds 100 ft, and individual coal seams can be more than 30 ft thick. Looking at the origin of the coals, it is originated in peat swamps located landward (southwest) of northwest/southeast trending shoreline sandstones. The location of the thickest coals coincides with the occurrence of high gas content, high rank, and high permeability in the play fairway.

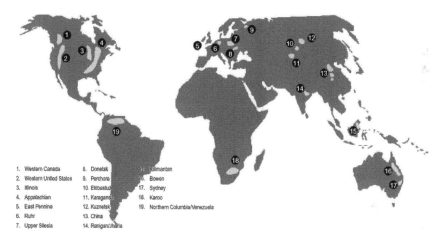

1. Western Canada	8. Donetsk	15. Kalimantan
2. Western United States	9. Perchora	16. Bowen
3. Illinois	10. Ekibastuz	17. Sydney
4. Appalachian	11. Karaganda	18. Karoo
5. East Pennine	12. Kuznetsk	19. Northern Columbia/Venezuela
6. Ruhr	13. China	
7. Upper Silesia	14. Ranigan/Jharia	

FIGURE 6.12   Major coal basins of the world. Modified from Thakur (2016).

6.6.1.2   POWDER RIVER BASIN

The Powder River Basin is one of the main coal-bearing regions along the Rocky Mountains, stretching from northern New Mexico to central Montana. The basin covers approximately 28,500 square miles, with approximately one-half of this area underlain by producible coals. The producing coals of the Powder River Basin are contained within the Tongue River member of the Paleocene Fort Union formation. The coals individually range up to 30 m thick with a total coal thickness of up to 90 m. (Ayers, 2000). Typical productive depths are 75–300 m compared with other US basins. Powder River Basin coals are immature with low gas contents and high permeability. Gas contents vary from 23 to 70 scf/ton (Ayers, 2000).

6.6.2   *GLOBAL DISTRIBUTION AND RESOURCES*

Coal is the major component of fossil fuel containing nearly 90% of the fossil fuel energy, and it is the most abundant and economical resource in the world today. Over the past 200 years, it has played a vital role in the stability and growth of the world economy. The growing population of the world would demand 5–7.5×10^{20} J (IEA, 2015) of energy to live well in the next 20 years. It is, therefore, essential that coal's share in the energy mix should increase.

At present, coal deposits are widespread in 70 countries of the world, and it is a very affordable and reliable source of energy. The total proved, mineable reserve of coal exceeds 1 T ton to a depth of about 3300 ft (1000 m). Indicated reserves (mostly no mineable) to a depth of 10,000 ft (3000 m) range from 17 to 30 T ton. Besides the minable coal reserve, the vast deep-seated deposits of coal contain another source of energy, that is, CBM. It is almost like natural gas with a slightly lower (10–15%) calorific value. The vast deposits of coal (17–30 T tons) contain approximately 9000 Tcf of gas (i.e., CBM). The prominent coal basins with a CBM reserve includes the United States, Western Canada, United Kingdom, France, Germany, Poland, Czech Republic, Ukraine, Russia, China, Australia, India, and South Africa. These countries produce 90% of global coal production and nearly 100% of all CBM production. Since the economic depth limit for mining is around 3000 ft, only about 1 T ton of coal can be mined leaving a vast reserve of coal full of CBM unutilized. Vertical drilling with hydrofracking is the main technique used to extract gas at present. This works only up to 3000–3500 ft depth because of serious loss in permeability. Global distribution and countries leading in CBM reserves are provided in Figure 6.13.

6.7   OIL SANDS

Oil sands are a naturally occurring mixture of heavy oil, water, and sand. According to Canadian Association of Petroleum Producers, oil sands are defined as either loose sands or partially consolidated sandstone containing a mixture of dense and extremely viscous form of hydrocarbon technically termed as bitumen with clay and water (Fig. 6.14). The composition of oil sands or tar sands is a combination of sand (83%), bitumen (10%), water (4%), and clay (3%). Bitumen is a semisolid, tar-like mixture of extremely viscous form of hydrocarbons. These are also termed as bituminous sands or tar sands in literatures and originated in the same manner as conventional deposits of oil, but microorganisms break down the light fraction of the oil. Bitumen is recovered from oil sands using two methods: mining and in situ drilling. The method used depends on how deep the reserves are. Energy developers extract oil from tar sands by injecting hot steam, which heats the sands and makes the tar less viscous so that it can be pumped out. Because generating the steam consumes large amounts of energy, tar sands are economically viable only when oil prices are high.

FIGURE 6.13 CBM reserves and activity.

Source: Modified from Al-Jubori (2009).

6.7.1 MAJOR PRODUCING REGIONS

According to Tar Sands World website, oil sands deposits exist in 70 countries. Canada holds more oil sands than anyone else in the world with 73% of global estimated oil sands and approximately constitutes 2.4 trillion barrels of oil in-place. The United States has less than 2% of global oil sands resources; however, 43% of US crude oil imports came from Canada in 2015, and over 64% of Canadian production comes from oil sands.

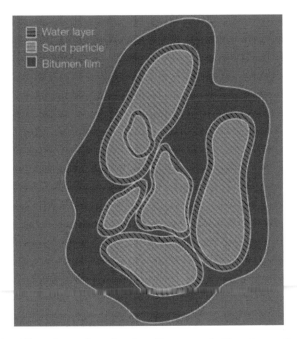

FIGURE 6.14   Oil sands—each grain of sand is surrounded by a layer of water and film of bitumen.

Source: Modified from Canadian Centre for Energy Information.

The world's largest oil sands deposit is widely distributed in Alberta of Canada. This is also known as Athabasca oil sands. This cretaceous oil sands deposit which covers an area of about 46,000 km² (Conly, 2002) is found in the McMurray Formation (a layer of shale, sandstone, and oil-impregnated sands formed by river and marine depositional processes). Stratigraphically, the McMurray Formation which is about 150 m thick, overlies the Devonian Waterways Formation (shale and limestone) and beneath the Clearwater Formation (a layer of marine shale and sandstone). The Clearwater Formation is itself overlain by the Grand Rapids Formation, which is dominated by sandstone (Conly, 2002).

The McMurray Formation is almost absent where the Devonian layer rises in a ridge, cutting through the McMurray Formation (Fig. 6.15). North of Fort McMurray, the formation can be found within 75 m of the surface and is exposed at the surface where the Athabasca River and its tributaries have incised into the landscape. The McMurray Formation is

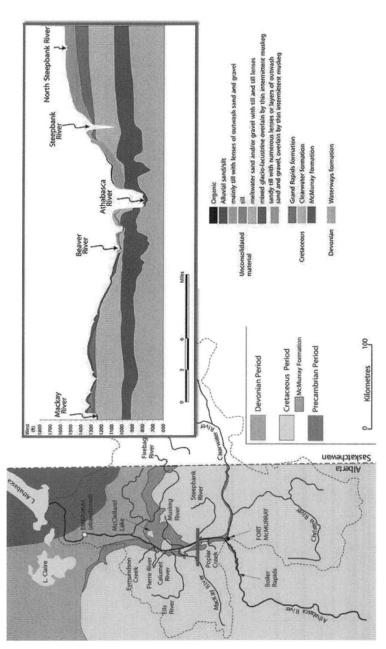

FIGURE 6.15 (**See color insert.**) Simplified geology of the McMurray Formation in the lower Athabasca region and an east–west cross-section map.

Source: Modified from Conly et al. (2002).

first exposed in the Athabasca riverbed at Boiler Rapids, 50 km upstream of Fort McMurray and again near the MacKay River (Conly et al., 2002).

The oil sand deposits, which occur less than 250 ft below the surface, are mined and processed to separate the bitumen. The bitumen is processed to produce synthetic crude oil before it is refined into petroleum products. The deeper occurring deposits are extracted by in situ (underground) methods, which include the likes of steam or solvent injection, or in situ combustion methods. The most common methods for recovery in Canadian regions are the cyclic steam stimulation and the steam-assisted gravity drainage (SAGD).

6.7.2   GLOBAL DISTRIBUTION AND RESOURCE ESTIMATES

Canada and Venezuela are hosts to the largest oil sand deposits in the world which occur in strikingly similar geological settings. These deposits are vast and comprise extra heavy (bituminous) sand with API above 20° hosted in unconsolidated shallow sandstones. Unconsolidated sandstones in the context of oil sand despots are characterized by high porosity, low tensile strength, and the absence of significant cohesion between the sand grains particularly because of oversaturation of oil which prevents consolidation into hard sandstone.

Oil sand resources in Canada and Venezuela together are of the order of 3.5–4 trillion barrels of original oil in-place (OOIP). Converted figures of reserves from this resource depend on the technological evolution of extraction and production methodologies which currently stand at about 10%. Technological developments in the late 1990s have evolved methods like SAGD using which a much greater percentage of the OOIP can be extracted compared to conventional methods. Using the current technological capability of 10% recovery, the Canadian reserves stand at about 200 billion barrels oil, whereas the Venezuela estimates are higher at 267 billion barrels. These figures make oil reserves of Canada and Venezuela comparable to the likes of Saudi Arabia, making these three countries own the largest oil reserves of the world. The total volume of unconventional recoverable oil from oil sands of these countries exceeds the reserves of conventional oil reserves of all the other countries put together. Should the recovery be enhanced from the current 10%–20% or higher, oil sands from Canada alone can quench the demand of North America for the next century. Although the oil reserves are plentiful, but being an unconventional resource, it is not a cheap oil and the challenge lies in advancement in technology to make the extraction and recovery more economic and environment friendly.

Oil sand in the United States are primarily concentrated in Eastern Utah where the known oil potential is 32 billion barrels from eight major deposits which include the ones in Carbon, Garfield, Grand, Uintah, and Wayne counties. These deposits, unlike the ones in Venezuela and Canada, are much smaller in size and occur in oil wet settings which are more difficult to extract and requires different extraction techniques.

Of the numerous other oil sand deposits, after Canada and Venezuela, the significant ones are the ones in Kazakhstan and Russia. In the Russian region, Eastern Siberia hosts oil sand deposits. The Tunguska Basin has the largest deposits at Olenek and Siligir. Other basins which contain oil sands are the Timan-Pechora and Volga-Urals basins. These basins are mature in terms of conventional oil resources in the shallow Permian formations. The North Caspian Basin, in Kazakhstan, is host to large bituminous sand deposits. Tsimiroro and Bemolanga are two heavy oil sands deposits of Madagascar. Tsimiroro currently has a pilot well and larger scale exploitation plans of the deposit is in early phase. Estimated reserves for oil sands in the Republic of the Congo are between 0.5 and 2.5 barrels.

6.8   CONCLUSION

The global natural oil and gas market is undergoing a major transformation driven by new supplies coming from the unconventional fuels. To meet the growing demand in developing economies, there is constant pressure to develop these unconventional fuels. This evolution of the role of unconventional fuels in the global energy mix has far-reaching consequences on energy trade as well as the security of global energy supplies.

Among the major sources of future oil and gas supplies, the unconventional fuels largely exceeds that of the other two sources, that is, development of known fields and yet to find sources. The wide occurrences of unconventional fuels are also known as continuous type deposits. The in-place volumes of unconventional fuels are estimated to be much higher than recoverable conventional oil and gas from known fields and yet to find sources. Unconventional hydrocarbons are found in low permeability, low porosity, and tight rock formations with low recovery, such as tight sands, shale gas, and oil and coal beds. The technological breakthroughs in horizontal drilling, directional drilling, and hydrofracturing have made shale gas and shale oil commercially viable in the United States, oil sandstone in Canada, and heavy oil in Venezuela. The world has already witnessed a significant increase in global unconventional petroleum production in the last decade.

KEYWORDS

- **shale gas**
- **shale oil**
- **tight sand**
- **hydraulic fracturing**
- **permeability**

REFERENCES

Al-Jubori, A.; Johnston, S.; Boyer, C.; Lambert, S. W.; Bustos, O. A.; Pashin, J. C.; Wray, A. Coalbed Methane: Clean Energy for the World. *Oil Field Rev.* **2009,** *21* (2), 4–13.

Ayers W. B., Jr. Coalbed Methane in the Fruitland Formation, San Juan Basin, Western United States: A Giant Unconventional Gas Play. In *Giant Oil and Gas Fields of the Decade 1990–1999*; Halbouty, M. T., Ed.; AAPG Memoir 78; 2003; pp 159–188.

Ayers, W. B. Methane Production from Thermally Immature Coal, Fort Union Formation, Powder River Basin. Paper Presented at the 2000 American Association of Petroleum Geologists Annual Meeting, New Orleans, 16–19 April.

Conly, F. M.; Crosley, R. W.; Headley, J. V. Characterizing Sediment Sources and Natural Hydrocarbon Inputs in the Lower Athabasca River, Canada. *J. Environ. Eng. Sci.* **2002,** *1* (3), 187–199.

Curtis, J. B. Fractured Shale-gas Systems. *AAPG Bull.* **2002,** *86*, 1921–1938.

David Waldo. A Review of Three North American Shale Plays: Learnings from Shale Gas Exploration in the Americas. Search and Discovery Article #80214 (2012). Posted May 28, 2012.

Dutton, S.; Clift, S.; Hamilton, D. *Major Low-Permeability Sandstone Gas Reservoirs in the Continental United States*; Bureau of Economic Geology, University of Texas: Austin, 1993.

EIA (US Energy Information Agency). *Review of Energy Resources: U.S. Shale Gas and Shale Oil Plays*; US Department of Energy: Washington DC, 2011.

Holditch, S. A.; Perry, K.; Lee J. *Unconventional Gas Reservoirs–Tight Gas, Coal Seams, and Shales*. Working Document of the National Petroleum Council on Global Oil and Gas Study, 2007.

IEA. *International Energy Agency Report*, 2015.

Jackson, J. A.; Bates, R. L. *Glossary of Geology*, 4th ed.; American Geological Institute: Alexandria, 1997.

Jarvie, D. M.; Hill, R. J.; Ruble, T. E.; Pollastro, R. M. Unconventional Shale-Gas Systems: The Mississippian Barnett Shale of North-Central Texas as One Model for Thermogenic Shale-gas Assessment. *AAPG Bull.* **2007,** *91* (4), 475–499.

Javadpour, F.; Fisher, D.; Unsworth, M. Nanoscale Gas Flow in Shale Gas Sediments. *J. Can. Petrol. Technol.* **2007,** *46* (10), 55–61.

King, G. E. Thirty Years of Gas Shale Fracturing: What Have We Learned? Presented at the SPE Annual Technical Conference and Exhibition, Florence, Italy, 19–22 September 2010; Paper SPE 133456.

Kumar, T.; Shandilya, A. *Tight Reservoirs: An Overview in Indian Context*. Presented at the 10th Biennial International Conference and Exposition, 2014.

Lash, G. G. Stratigraphy and Fracture History of the Middle and Upper Devonian Succession, Western New York: Significance to Basin Evolution and Hydrocarbon Exploration. In *Pittsburgh Association of Petroleum Geologists Spring Field Trip Guidebook*; Pittsburgh Association of Petroleum Geologists: New York, 2008; p 88.

Law, B. E.; Curtis, J. B. Introduction to Unconventional Petroleum Systems. *AAPG Bull.* **2002,** *86*, 1851–1852.

Misra, R.; Dasgupta, D. K. In *Shale Gas Indian Strides*, Proceedings of the Indian National Science Academy, 78, No. 3, September 2012.

Naik, G. C. Tight Gas Reservoirs—An Unconventional Natural Energy Source for the Future, 2005.

Rightmire, C. T.; Eddy, G. E.; Kirr, J. N., Eds. *Coalbed Methane Resources of the United States*. American Association of Petroleum Geologists: Oklahoma, 1984.

Schmid, K. W., and Markowski, A. K., 2017, Source rock evaluation of the Upper Devonian Genesee, Harrell, and West Falls Formations in Pennsylvania: Pennsylvania Geological Survey, 4th ser., Mineral Resource Report 102, 45 p.

Shtepani, E.; Shtepani, L. A.; Noll, L. W. Elrod; Jacobs, P. M. A New Regression-based Method for Accurate Measurement of Coal and Shale Gas Content. *SPE Reserv. Eval. Eng.* **2010,** 359–364.

Thakur, P. *Advanced Reservoir and Production Engineering for Coal Bed Methane*. Gulf Professional Publishing: West Virginia, 2016.

US Geological Survey's Energy Resources, 2016.

Zhang, T. W.; Ellis, G. S.; Ruppel, S. C.; Milliken, K.; Yang, R. S. Effect of Organic Matter Type and Thermal Maturity on Methane Adsorption in Shale-gas Systems. *Org. Geochem.* **2012,** *47*, 120–131. DOI:10.1016 /j.orggeochem.2012.03.012.

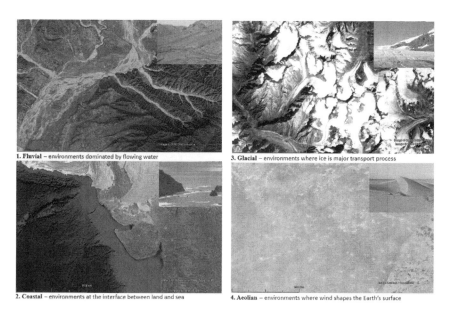

1. Fluvial – environments dominated by flowing water

3. Glacial – environments where ice is major transport process

2. Coastal – environments at the interface between land and sea

4. Aeolian – environments where wind shapes the Earth's surface

FIGURE 1.7 Erosional and depositional environments.

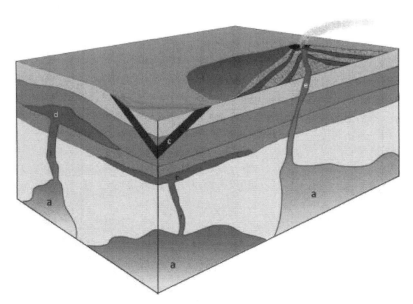

FIGURE 2.3 Different types of the plutons: (a) stocks, (b) sill, (c) dyke, (d) laccolith, (e) pipe, and (f) pipes or dykes.

Source: Reprinted from Earle (2014). https://opentextbc.ca/geology/chapter/3-5-intrusive-igneous-bodies/ https://creativecommons.org/licenses/by/4.0/

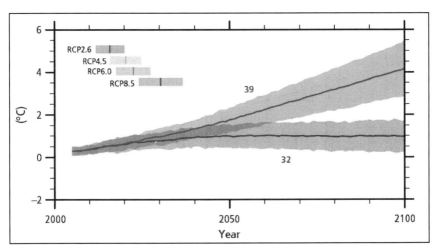

FIGURE 3.7 RCP scenario-based global surface temperature change predictions.

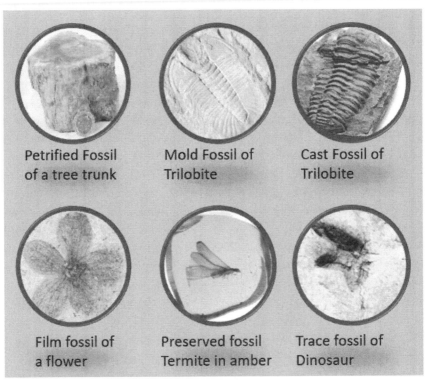

FIGURE 4.2 Types of fossils found in rock records.

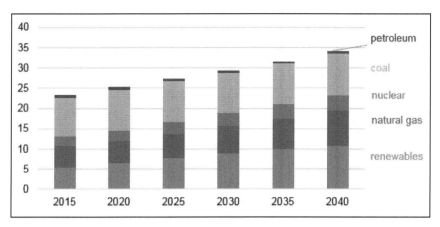

FIGURE 5.2 World net electricity generation by energy source (in trillion kilowatt-hour).

Source: International Energy Outlook (2017). (https://www.eia.gov/outlooks/ieo/pdf/exec_summ.pdf)

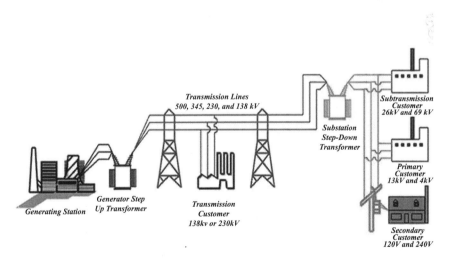

FIGURE 5.10 Basic structure of the electric system. Blue lines are transmission lines, green are distribution, and black lines represent power generation lines.

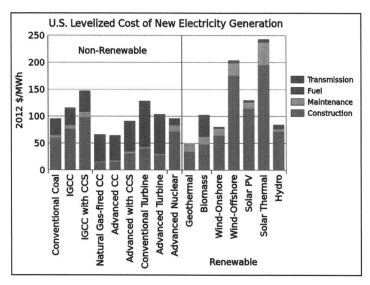

FIGURE 5.12 Estimated levelized cost of electricity generation (in $/megawatt-hour) in 2018 based on 2011 statistics.

Source: Delphi234. (https://commons.wikimedia.org/wiki/File:LCOE_US_Plants_Coming_Online_in_2019.svg)

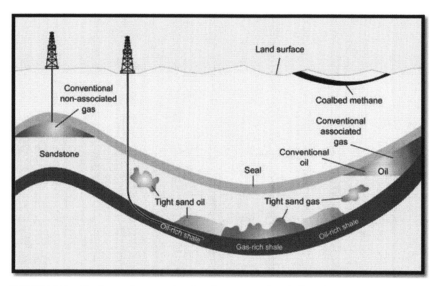

FIGURE 6.2 Geology of conventional and unconventional oil and gas.

Source: US Energy Information Administration. (https://www.eia.gov/todayinenergy/detail.php?id=110)

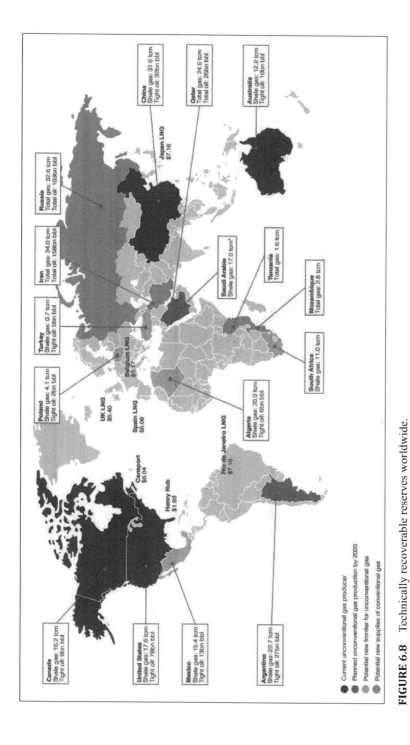

FIGURE 6.8　Technically recoverable reserves worldwide.

Source: Reprinted with permission from World Energy Council, 2016.

FIGURE 6.11   The world map representing unconventional gas reserves in various regions.

Source: Pacwest Consulting.

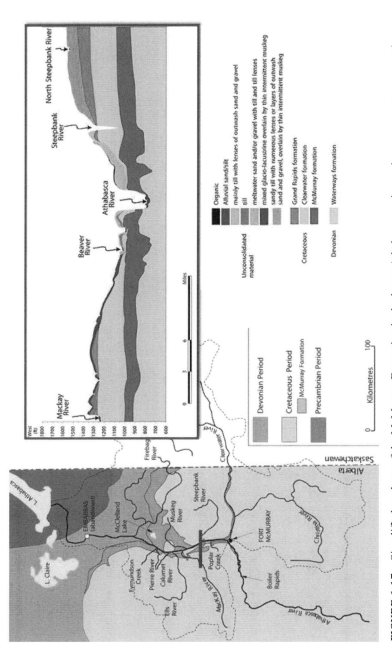

FIGURE 6.15   Simplified geology of the McMurray Formation in the lower Athabasca region and an east–west cross-section map.

Source: Modified from Conly et al. (2002).

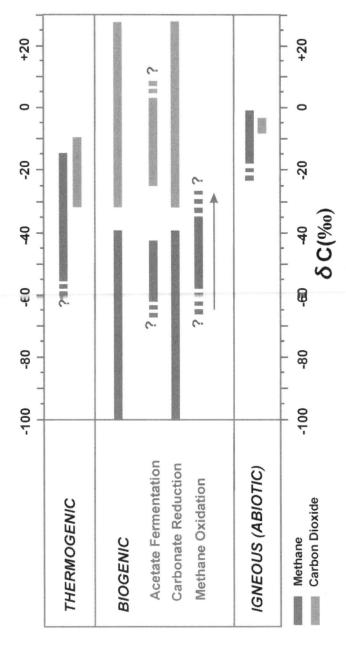

FIGURE 7.2   Range of isotopic fraction of C in different modes of formation.

Source: Modified from Whiticar, 1990.

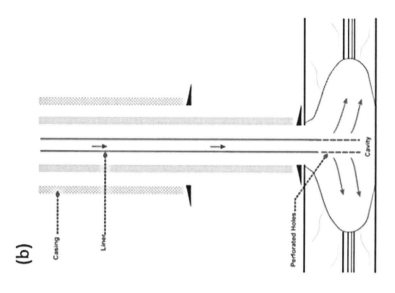

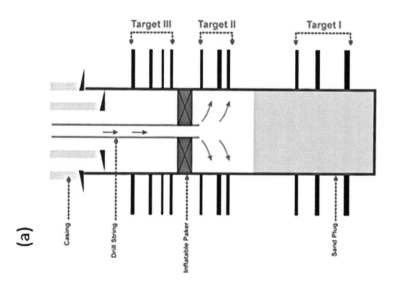

FIGURE 7.7 (a) Open-hole completion technique. (b) Cavity completion technique for thick seam.

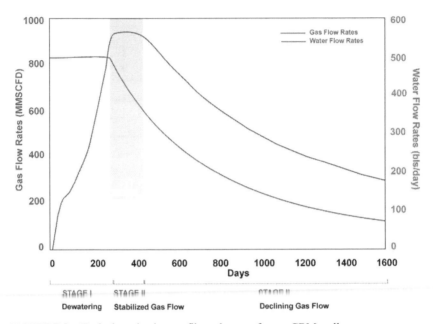

FIGURE 7.8   Typical production profile and stages from a CBM well.

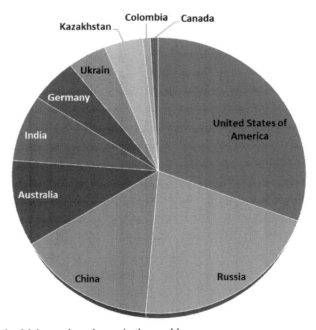

FIGURE 8.1   Major coal producers in the world.

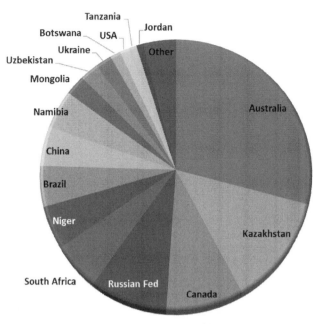

FIGURE 8.2   Major producers of uranium in the world.

FIGURE 8.11   A schematic presentation of a geological model of a coal deposit.

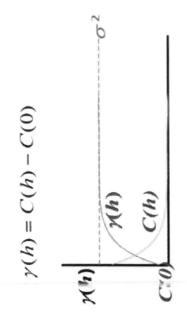

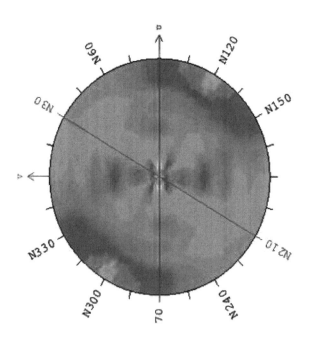

FIGURE 8.19   A variogram map (left) and a variogram model (right).

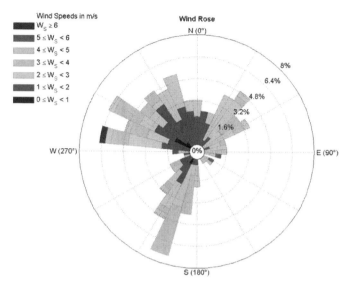

FIGURE 9.5 Wind rose diagram for Changi region Singapore.

Source: prepared with data downloaded from www.windfinder.com.

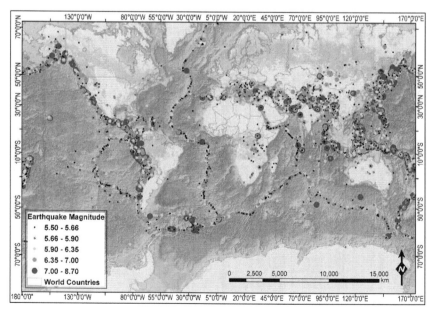

FIGURE 11.7 Historical earthquakes worldwide from 1965 to 2016.

Source: Map generated from data sourced from National Earthquake Information Center.

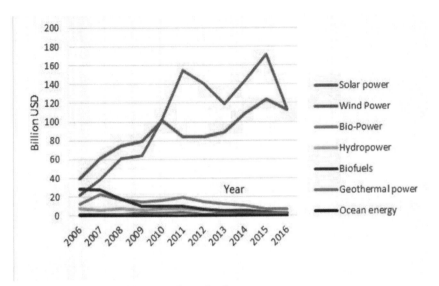

FIGURE 12.2   Renewable energy by technology.

Source: Reprinted with permission from REN21, 2017.

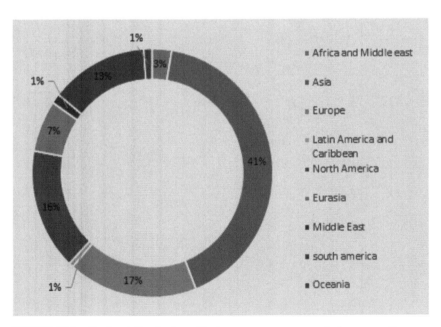

FIGURE 12.5   Regional distribution of hydropower energy capacity in megawatts.

Source: Reprinted with permission from REN21, 2017.

FIGURE 12.10 Global wind power capacity in W/m².

Source: National Academy of Sciences of the United States of America. (https://www.pnas.org/content/106/27/10933)

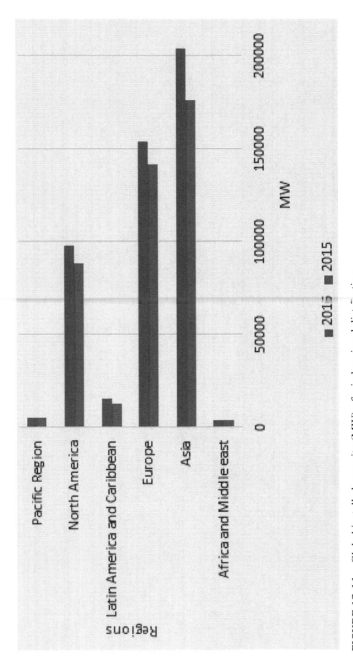

FIGURE 12.11   Global installed capacity (MW) of wind regional distribution.

Source: Reprinted with permission from REN21, 2017.

CHAPTER 7

Coal Seam Gas: Evaluation, Extraction, and Environmental Issues

DEBADUTTA MOHANTY*

Nonconeventional Gases Department, CSIR-Central Institute of Mining and Fuel Research, Dhanbad 826015, India

*E-mail: drdmohanty@ymail.com

ABSTRACT

Coal acts as both source rock and reservoir rock for methane. This methane extracted from coal seams is known as coal seam gas. Methane is a common source of energy used in modern society and at the same time it is an environmental hazard, as a combustible gas in underground coal mines and as a greenhouse gas in the atmosphere. Hence, its evaluation, extraction, and related environmental issues are important. Methane remains mostly as adsorbed molecules on micropores in coal. "Direct method" using a desorption canister is used for determining the gas content of coal and its storage capacity is determined through adsorption isotherm construction. Coal seam gas extraction involves drilling, well completion, and well stimulation. Main environmental concern is the disposal of the co-produced water which has high soda content. The chapter provides a synoptic view on various evaluation and extraction techniques of coal seam gas with a brief on clean coal technologies which addresses the associated environmental concerns of coal usage.

7.1 INTRODUCTION

Coal acts as a reservoir of significant amounts of methane and other gases in traces which are collectively known as coal seam gas (CSG) or coalbed methane (CBM). Unlike conventional oil/gas reservoirs, coal acts as the source, the reservoir, and the trap for methane. CSG has grown from an unconventional gas play, which most operators deemed uncommercial some 30 years ago into a commercially profitable and sustainable source of natural gas. Most of the gases are stored in the coal seams are in the adsorbed state which is very different from the way conventional gas is stored in porous and permeable reservoirs. The adsorption of gases on coal surface is due to weak molecular attractions resulting from Van der Waal forces. A fraction of the gas occurs dissolved in water within the coal matrix. The gas storage capacity of coal is extremely high and in multiples times as that of conventional reservoirs at low pressures.

Mining of coal releases methane into the atmosphere. Being a potential hazardous gas, the release of methane into the environment can cause explosion and outburst. Being a greenhouse gas its increased levels in the atmosphere contributes to global warming and climate change. Methane has 21 times more the global warming potential than carbon dioxide. It is therefore important that this gas is extracted and put to energy production in an environmentally friendly manner before mining of coal. Methane is a clean source of energy when compares to coal. All coal seams do not possess the same gas potential and therefore evaluation of methane content is a prerequisite for its production and recovery as coal mine methane (CMM) coal bed methane (CBM). Proper estimation of methane content of coal seams is a statutory requirement for opening a mine, as well as it is highly imperative for designing the ventilation circuits for underground mines.

7.2 COAL SEAM GAS

7.2.1 BIOGENIC VERSUS THERMOGENIC GAS

The gas in coal originated and is stored in the coal itself. Methane is generated naturally from organic matter through microbial (biogenic) or thermal (thermogenic) processes which are triggered alongside sediment burial and initiation of diagenesis. Diagenesis of humus forms coal and gas simultaneously and the gas is physically sorbed in the coal itself. Coal being brittle is fractured under tectonic stress which increases its surface areas aiding adsorption of methane. A gram of coal can have a surface area as large as a football field and therefore is capable of sorbing large quantities of methane. A ton of coal can produce about 1300 m^3 of methane. The different gases generated during different stages of coalification are shown in Table 7.1. Hydrocarbon generation primarily depends on quantity, type, and structure of the kerogens and is controlled by temperature and pressure to which the precursor material is subjected over the geologic time.

TABLE 7.1 Types of Coal Gases.

Gas-generating stage	Random, Rr (%)
Primary biogenic methane	<0.30
Early thermogenic methane	0.50–0.80
Maximum wet gas generation	0.60–0.80
Onset of main-stage thermogenic methane	0.80–1.00
Onset of secondary cracking of condensate	1.00–1.35
Maximum thermogenic gas generation	1.20–2.00
Deadline of significant wet gas generation	1.80
Deadline of significant methane generation	3.00
Secondary biogenic gases	0.30 to 1.50+

Source: Adapted from (Scott, 1993).

Methane (CH_4) is generated in significant quantities in sediments, at both near-surface conditions (by bacterial decomposition) and at greater depths (by thermogenic cracking). The biogenic stage methane may be produced by bacterial activity after reaching the thermal maturity. In the early stage of coalification, methane is generated at shallow depths (<10 m) in lower-rank coals (subbituminous) up to a temperature of about 50°C and is termed as "biogenic" or "diagenetic methane." The biogenic gases formed during early stages of coalification contributes only 10% of the total methane generated, most of which escapes into the atmosphere due to exposed peat surface and low degree of confinement at surface conditions. But some amount may store under certain specific geologic conditions like rapid subsidence and burial. Thermogenic gases, however, formed in higher rank coals at a later stage of coalification and greater depths, and the bulk of this methane is stored in coal seams due to a higher degree of confinement. The gas generation by thermogenic process begins at vitrinite reflectance (Rmax) of 0.70–0.80%.

The degree of hydrocarbon generation is conventionally described by the kerogen types based on H/C versus O/C diagram. The generalized diagram of the generation of biogenic and thermogenic methane with rank enhancement of coal has been shown in Figure 7.1. Biogenic methane is produced due to microbial activity with the sediment accumulation rates of about 50 m/million years. The processes of formation of thermogenic gases (C1–C4) are slower than that of biogenic gases. Generally, high concentrations of thermogenic gases in sediments at shallow depths and low temperatures indicate the existence of hydrocarbon migration pathways in case of conventional reservoirs. C2–C4 gases of lower concentration also may form along with C1 because of bacterial activity but migration generally does not take place in coal reservoirs.

Biogenic and thermogenic gases usually distinguished on the basis of carbon isotopic ratios. Isotope fractionation is the natural partitioning of isotopes because of the subtle differences (primarily in weight) among isotopes. Bonds are easier to break in "lighter" isotopes, indicating that these isotopes are more chemically reactive. The fractionation factor quantifies the amount of separation that occurs among isotopes of the same type atom at any given temperature; more fractionation (separation of lighter and heavier isotopes) occurs at lower temperatures. Isotopic fractionation is generally expressed as delta (δ) symbol:

$$\delta^{13}C = (R_A - R_{STD})/R_{STD}$$
$$R = {}^{13}C/{}^{12}C.$$

C isotope values are relative to the PDB standard (Pee Dee Belemnite). H (Deuterium, δ^2H or δD) and Oxygen isotopes are relative to standard mean ocean water or SMOW. Isotopic values are reported in parts per thousand or per mil, for example, a $\delta^{13}C$ isotopic value of +55‰ indicates that the sample is enriched in the ^{13}C isotope by 55 parts per thousand.

Both microbial CO_2 reduction and fermentation (aceticlastic) reactions can occur during microbial activity in the coals; thus, both carbon and hydrogen isotopes are necessary to distinguish between these two methanogenic pathways (Flores et al., 2008). Carbon and hydrogen isotope ratios in natural methane range widely between −50 and −110‰ for $\delta^{13}C$ and from −150 to −400% for δD. Microbial and fermentation processes cause depletion or enrichment of various isotopes in methane. Microbial CO_2 reduction causes ^{13}C depletion which results in $\delta^{13}C$ ranges between −55 and −110%. Whereas microbial methyl type fermentation enriches ^{13}C with $\delta^{13}C$ between −40 and −70%. Methane generated from microbial CO_2 reduction generally has higher δD values (−150 to −250‰) as compared to methane from microbial methyl-type fermentation where δD values range between −250 and −400%. Figure 7.2 shows the range of an isotopic fraction of C in different modes of formation.

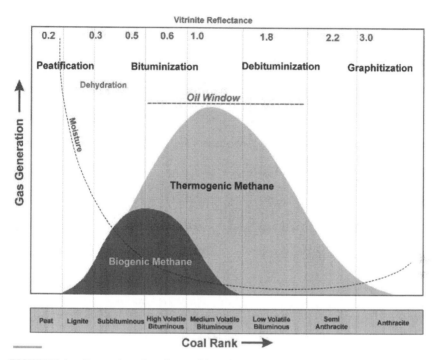

FIGURE 7.1   Generation of methane with rank enhancement of coal.

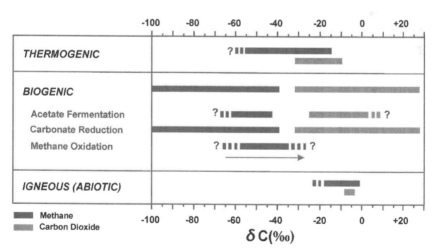

FIGURE 7.2   **(See color insert.)** Range of isotopic fraction of C in different modes of formation.

Source: Modified from Whiticar, 1990.

7.2.2 STRUCTURE OF COAL SEAMS

Physical sorption is dependent on the nature of the substrate, that is, the porous structure and composition of coal and the degree of confinement. The matrix of coal is heterogeneous and displays porosity anisotropy. Porosity in coal can result from matric porosity (macro-, meso-, and micropores) or from induced porosity from fractures and cleats. IUPAC nomenclature for pore size is as mentioned below:

Macropore: Pore of width $w \geq 50$ nm
Mesopore: Pore of width 50 nm $\leq w \leq 2$ nm
Micropore: Pore of width $w \leq 2$ nm

Micro pores are the dominant site for methane adsorption surface. Regional and local structural events and de-volatization stage of coalification results in the formation of fractures called cleats. Cleat patterns are scale dependent and hierarchical. Small-scale micro-fractures that develop within the coal seam are called secondary and tertiary cleats. Primary cleats are the ones which cuts across coal layers into non-coal formations between seams and are often referred to as master cleats. Cleat patterns are crucial for gas production because they allow passage for the releases gas which was sorbed in the coal seams. Deep burial of coal can destroy cleats and the process is called healing which is aided by secondary mineralization. Some characteristics of cleats include:

- Interconnected network in the coal appearing perpendicular to bedding
- Butt cleats are shorter and terminate at a face cleat
- Face cleats and butt cleats are mutually perpendicular
- Normal spacing between cleats are less than 25 mm
- Intrinsic tensile force and associated shrinkage, fluid pressure and tectonic stress are common causes of formation of cleats
- The geometric pattern of the cleats is tectonic stress controlled (Fig. 7.3a)
- Permeability along face cleats can be up to 17 times higher than along butt cleats (Fig. 7.3b)

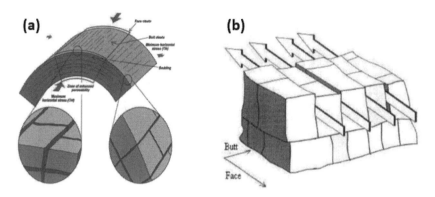

FIGURE 7.3 (a) Effect of tectonic stress on cleat development. (b) Variation in permeability with cleat type.

7.2.3 GAS RETENTION AND RELEASE MECHANISM

The gas retention mechanism in coal seams is markedly different from those of conventional gas reservoirs. Methane remains as adsorbed molecules on surface, trapped gas within matrix, free gas in cleats and fractures, and as a solute in ground water in coalbeds (Fig. 7.4a). Physical adsorption is the dominant mechanism governing the storage of CBM in the coal seam. Adsorption phenomenon is caused by weak molecular attractions and the gas is ultimately held in the coal hydrostatically. In adsorption, the adsorbed species is the adsorbate and the material adsorbing the species is the adsorbent. In adsorption processes atoms, molecules or ions in a gas or liquid phase diffuse to the surface of an adsorbent and bind to the surface as a result of London—Van der Waals forces and/or nondispersion forces. IUPAC extended the original adsorption isotherm classification proposed by Brunauer, Deming, Deming, and Teller to give six generic adsorption isotherms. Type I or Langmuir isotherm (Fig. 7.4b) is of immense use in CBM production engineering. The main parameters influencing the gas adsorption characteristics of a coal seam are coal type and rank, moisture content, temperature, in situ stress, nature of mineralization, fracture development, and the degree of confinement.

CSG/CBM is generally not free-flowing to the well-bore. During extraction of methane, the reservoir pressure is reduced by dewatering the coal seams. Gas production from coal is a complex process whereby gas, initially adsorbed in the coal matrix, desorbs and diffuses through the matrix into the cleat and eventually flows through the cleat system into a

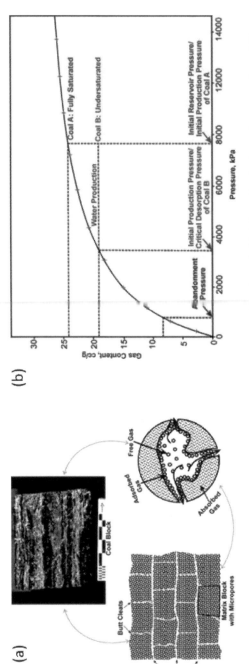

FIGURE 7.4 (a) Porous structure and different modes of occurrence of methane in coal. (b) Adsorption isotherm for reservoir modeling.

production well or a drainage borehole (Fig. 7.5). Fluid transport through the reservoir is mainly controlled by the permeability of coal seams. The permeability of coal is generally very low, varying between 0.5 and 100 md, and mostly contributed by the interconnected network of fractures. Hence, permeability is the most important parameter for injection/recovery of gas into/from coal seams.

Permeability and porosity are the main determining factor for the flow characteristics of the coal reservoir gases. As it generally contains limited natural fractures, extraction of gas is not possible without hydraulic fracturing. The amount of methane in the coalbed usually increases with an increase in surface area of organic matter and/or clays. The gas that resides in fractures and pores can be easily released than the adsorbed gas and is called free gas. The higher occurrence of free-gas in coal generally results in initial higher production rates which in about a year of production declines rapidly to a low but steady rate of production when the adsorbed gas is slowly released. On a unit concentration basis, CO_2 causes a greater degree of coal matrix swelling compared to CH_4 due to differing sorption capacity of coal toward these adsorptive gases. This phenomenon is used for enhanced recovery of CBM from depleting wells through simultaneous injection of CO_2.

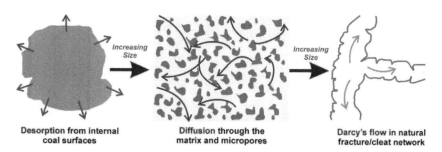

| Desorption from internal coal surfaces | Diffusion through the matrix and micropores | Darcy's flow in natural fracture/cleat network |

FIGURE 7.5 Schematic representation of gas flow through coalbeds during CBM production.

7.3   GAS EVALUATION

7.3.1   *IN SITU GAS ESTIMATION*

Of the various methods for determination of in situ gas content in a sample of coal, the "Direct" method is accepted and practiced widely for the estimation of coal bed methane resource potential in a virgin coal blocks.

This method involves the collection of coal core samples/coal sample, estimation of lost gas (Q_1), measurement of desorbed gas (Q_2), determination of residual gas (Q_3) and calculation of total gas volume. Coal cores are collected during the exploratory drilling of boreholes in a cylindrical canister for the measurement of desorbed gas volume. When coal samples are recovered from a welbore, some volume of gas will desorb from the samples before sealing it in a desoption canister. This amount of gas is known as lost gas (Q_1) and it is estimated on the basis of lost time. The amount of lost gas depends on lost time, which is elapsed during retrieval and sealing in the canister. Time of coal seam encountered the start of coal cutting, its retrieval and lapsed time until the sample is sealed in canister must be recorded to calculate the lost gas volume. The lost gas can be calculated by graphical method based on the relationship that the volume of gas released for the first few hours is proportional to the square root of the desorption time (Fig. 7.6a).

After an extended period of desoption measurement, the coal sample will effectively cease to release the adsorbed gas (Q_2). Desorbed gas volume is the portion of the total gas released from a coal sample sealed in a desorption canister. The volume of desorbed gas is periodically measured at ambient conditions by allowing the positive pressure in the canister to bleed in an appropriate manometric arrangement. The volume of gas released is measured as a function of time. Gas measurement will continue until the volume of gas released, typically falls below 0.05 cc/g per day. The desorbed gas volumes need to be corrected for standard temperature and pressure conditions. Cumulative values obtained by desorbed gas volume measurement gives the desorbed gas (Q_2) for a coal sample.

Any gas still present in the sample is called residual gas (Q_3). Residual gas may be largely unrecoverable in coalbed methane production. Subsequent to determination of Q_2, the core sample is crushed in an airtight steel vessel, initially purged of air prior to crushing, to fine powder below 200 mesh BSS size. The volume of residual gas released on crushing is measured by the manometric method and the residual gas (Q_3) is calculated by unitary method for the total weight of the sample.

Gas (methane) content is calculated from the total volume of gas obtained by the addition of Q_1, Q_2, and Q_3 divided by total weight of the sample.

Gas content (cc/g), $Q = (Q_1 + Q_2 + Q_3) / W$

where, W = total weight of sample

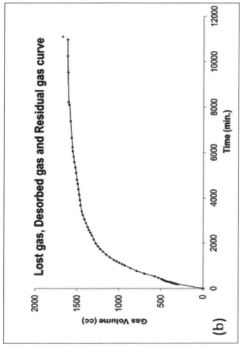

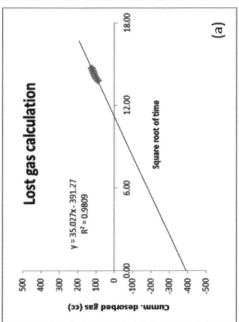

FIGURE 7.6 (a) Lost gas (Q_1) calculation. (b) Gas desorption curve.

7.3.2   ADSORPTION ISOTHERM

Methane sorption capacity of coal is the amount of gas that can be adsorbed at a particular pressure and is determined through adsorption isotherm construction. Mainly there are two techniques to determine the sorption capacities of gases on coals, namely, volumetric (successive gas expansions into an accurately known volume containing the adsorbent) and gravimetric (gas expansions into the enclosed microbalance containing the adsorbents) methods. These are different in terms of their experimental setup, devices, and operational conditions and require a long time to achieve equilibrium at various pressure steps.

   The experiment involves the introduction of a known quantity of adsorbent (CH_4/CO_2) to an adsorbate (coal/shale) and the amount of gas adsorbed by the coal/shale can be calculated by measuring the drop in the pressure of the system. The amount and size fraction of the sample depends upon the design of the experimental setup and objective of the investigation. However, for volumetric adsorption isotherm experiments the sample amount is typically between 100 and 150 g in the grain size fractions of −0.25 mm or 60 BSS mesh and −1 mm or 16 BSS mesh sizes. Moisture content is an important influencing parameter in adsorption isotherm study. It is generally found that the coal beds are saturated with moisture under reservoir conditions. Hence, moisture equilibration is required to maintain a standard condition for analysis and provides a means to determine the storage capacity of the coal seams at reservoir conditions. For this reason, moisture equilibration is required prior to the adsorption isotherm experiment. ASTM D1412/1412M standard is followed for the equilibrium moisture of coal at 96% relative humidity (RH) at 30°C.

7.3.2.1   ADSORPTION ISOTHERM CONSTRUCTION

Isotherm is commonly used to present adsorption equilibrium. At a constant temperature, a plot of adsorption capacity versus adsorption equilibrium pressure represents adsorption isotherm. The isotherm construction is important to analyze the maximum monolayer adsorption capacity, adsorption type and porous structure of adsorbate. A detailed study on adsorption isotherm provides information about the saturation level of coal, critical desorption pressure, the abandonment pressure of the reservoir, production profile and life of the well, calculation of total recoverable gas from the well, to evaluate the economic viability of the CBM block, and so on. A generalized adsorption isotherm curve is shown in Figure 7.6b.

The adsorption isotherm experiment involves mainly two steps, that is, (1) Measurement of dead/void volume within the experimental setup. (2) Determination of adsorption capacity.

The volume which is not occupied by the sample is known as dead volume or void volume. To determine the void volume, a nonadsorbing gas like helium is used. Being an inert gas and having the smallest molecule size, helium can easily permeate into the micropores in coal. The void volume is the volume available to the gas-phase in the experimental setup, which is not occupied by the skeletal volume of the solid adsorbent. After loading the sample into the sample cell, helium expansion from reference cell into sample cell is performed in order to determine the void volume. The ratio of initial volume (reference cell volume) to that of final volume (reference cell volume plus volume available within the sample cell) intruded by helium can be obtained by mass balancing.

After the dead volume is determined, the entire setup is evacuated for adsorption isotherm construction. Methane is added to the reference cell and allowed to equilibrate at the bath temperature before admitting it to sample cell. The sample cell is then connected to reference cell and allowed to equilibrate. The drop in pressure so caused is corrected for void volume to calculate the actual amount of gas adsorbed in coal. The procedure is repeated at increasing pressure steps of about 0.5–1.0 MPa until the desired highest pressure is achieved. The volume of gas adsorbed at each pressure step is used to construct the adsorption isotherm.

Adsorption isotherm calculation is based real gas law that considers the compressibility factor and the experimental values are fitted to the Langmuir equation for determination of maximum monolayer adsorption capacity. Langmuir equation relates the area covered by the adsorbate gas molecules on the surface of solid adsorbents as a function of its partial pressure or concentration at a fixed temperature and is based on the following assumptions:

- Molecules are adsorbed at discrete active sites on the surface.
- Each active site adsorbs only one molecule.
- The adsorbing surface is generally uniform.
- There is no interaction between the adsorbed molecules.

Hence, the rate of adsorption of adsorbate molecules at the surface is proportional to the fraction of the uncovered area $(1-\theta)$ and desorption rate is proportional to the surface covered, that is, θ. The rate of adsorption is $k_1(1-\theta)$ and rate of desorption is $k_2\theta$.

At equilibrium,

Rate of adsorption = Rate of desorption, that is, $k_1(1-\theta) = k_2\theta$

$$\text{or, } \frac{\theta}{\theta-1} = \frac{k_1}{k_2},$$

$$\text{or, } \theta = \frac{k_1 P}{k_2 + k_1 P}$$

$$\text{or, } \theta = \frac{K_L P}{1 + K_L P} \qquad \text{where } K_L = k_1/k_2.$$

Surface coverage can be written as the ratio of sorbed amount to the maximum amount that can be sorbed by the adsorbent, that is, $\theta = \frac{q}{q_m}$

Hence, the Langmuir equation is $q = q_m \dfrac{K_L P}{1 + K_L P}$ where K_L is the Langmiur constant.

At a very low partial pressure of the adsorbate, $K_L P \ll 1$, $1 + K_L P \approx 1$. The low-pressure region of the adsorption isotherm is known as Henry's law region and can be written as

$q = q_m K_H P$ where K_H is the Henry's law constant.

Langmuir equation can be written as,

$$q_m K_L P = q(1 + K_L P)$$

$$\text{or, } \frac{P}{q} = \frac{1 + K_L P}{q_m K_L}$$

$$\text{or, } \frac{P}{q} = \frac{1}{q_m K_L} + \frac{P}{q_m}$$

The amount of gas sorbed per unit increase in pressure decreases with increasing sorption pressure and the sorbed gas eventually reaches a maximum value which is represented by Langmuir volume constant (V_L) and the pressure at which gas storage capacity equals one half of the maximum storage capacity $(V_L/2)$, represents Langmuir pressure constant

(P_L). Langmuir parameters are calculated from the model fit for adsorption isotherm data. Langmuir volume and Langmuir pressure are critical for CBM reservoir characterization.

7.4 GAS EXTRACTION

Worldwide CBM plays show lateral and vertical variations even within the same basin. CBM industry involves both geological and technological challenges for commercial scale production. After CBM plays are carved out based on the knowledge of geological and reservoir complexities, appropriate drilling, completion, stimulation, and operation technologies are chosen to achieve significant rates of gas production. CBM technology evolved gradually through phases. For this purpose extensive data collection through coring, specialized well logging and in certain cases, special seismic techniques may be adopted.

7.4.1 CBM WELL DRILLING

CBM activities are generally carried out in several phases, that is, exploration, pilot tests, development, and production. The success of one phase leads to the next. In each phase, drilling practices evolves to suit the requirement of the phase. Drilling of CBM wells requires attention to the exhaustive reservoir data collected from the assessment core holes drilled during exploration phase (Phase-I). Vertical coal bed methane drainage wells are mainly used in the United States in Black Warrior Basin and San-Juan Basin though these basins differ considerably in their age, rock mechanical properties and permeability. Permeability is one of the deciding factors for drilling horizontal drilling. Low-permeability coals with thickness of 3 ft or more may require horizontal completions. Multilateral in-seam drilling technique for horizontal wells in unconventional reservoirs has proven successful. Underbalanced drilling (UBD) techniques are typically deployed to minimize formation damage. This involves applying more pressure in the formation through use of appropriate liquids with some solids and air to maintain the return pressure for controlling fluid influx in the overpressured basins.

In exploration phase, a well of 900–1000 m is completed using conventional drilling techniques in 30–40 days—whereas in assessment/

pilot phase drilling time is reduced to 20–25 days per well for drilling up to 600–900 m using conventional and UBD techniques. In the development phase, where the total depth (TD) of wells vary from 300–900 m is drastically reduced to 5–7 days by using air-hammer and air-foam drilling. Air drilling is done with air-hammer bits while fluid drilling is mostly done using tri-cone rotary bits. Air drilling or use of freshwater systems is economical and environmentally appealing over the use of drilling mud. Geomechanical properties of coal seam is critical in selection of appropriate drilling fluid as minimum use of surfactants, lost-circulation solids, polymers, etc. will reduce the risk of formation permeability damage. As most CBM wells are drilled with clear fluids, cleaning can be done effectively with higher pump rates.

Mainly conventional cased hole completion method is adopted with two-casing policy in all the phases. 8 5/8" surface casing is lowered in 10 5/8" hole whereas 5 ½" production casing is lowered in 7 7/8" hole. The major challenge in cementing CBM wells is to control fluid invasion into the delicate cleat system. Though the hole is drilled underbalanced, cementing job requires slightly overbalancing to prevent free-gas migration into the cement column after the casing is placed. Foam cement being lightweight puts less pressure on the delicate cleat structures, helps as lost-circulation plugs and provides durable object isolation for CBM wells.

7.4.2   COMPLETIONS

Coal is friable with extensive natural fracture system that needs to be connected to the wellbore to provide adequate permeability. Adsorptive properties of organic compounds in drill mud and frac fluids may lead to swelling of the coal matrix, thereby reduces the permeability. High treatment pressure during fracking and handling coal fines generated during production poses additional issues. Hence, CBM well completions need modifications over the conventional gas well completion techniques and evolved from open-hole completion technique through open hole cavitation process, cased-hole completions to multizone entry in cased hole.

The first type of completion used in the Warrior basin were open hole completions (Fig. 7.7a) where the prime aim was to recover gas from a single seam with minimum formation damage. This involves the setting of the casing above the coal and drilling through the coal followed by hydraulic fracturing of the seam. Then the well is cleaned with compressed

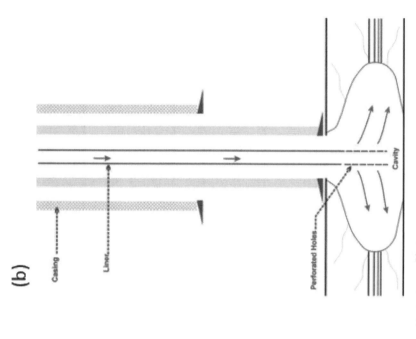

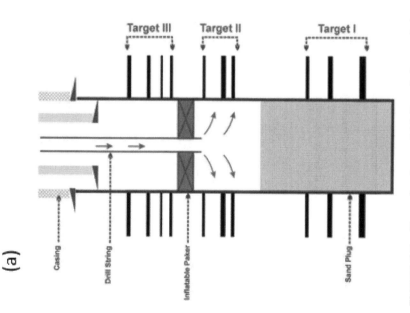

FIGURE 7.7 **(See color insert.)** (a) Open–hole completion technique. (b) Cavity completion technique for thick seam.

air and pumping equipment is lowered for production. Cavity completion technique (Fig. 7.7b) developed in open hole for thick and overpressured seams of high permeability to intersect natural fracture system to effectively connect the formation cleat network to the wellbore.

Coal mostly occurs in cyclothems (a cyclic pattern of deposition of sedimentary strata). Multiple groups of thin coal seams are therefore common. To exploit gas from multiple seams, multizone cased-hole completions are required where each seam is accessed by slotting or perforation. The slotted-casing technique used for better control on the fracturing job as compared to open hole completions. In this technique a large coal face is opened to the bore well while isolating each zone for controlling the fracture fluid entry. The procedure involves drilling, casing and cementing at different hole internals. A jetting tool, attached to the end of the tubing string, is used through the casing slops through which cement with high-velocity streams of an abrasive water–sand solution is jetted. For access the gas, the casing is first set to the total depth, and each zone is perforated at four shots per foot and fractured by jetting sand slurry. Perforating is inexpensive compared to slotting and saves more time in well completion.

7.4.3   WELL STIMULATION (FRACKING)

As the reservoir rock is dewatered, most of the sorbed gas diffuses to fractures where it is released and migrates to the wellbore. Fracture stimulation treatment is gradually becoming an important activity for coalbed methane productivity around the world. Since coalbed methane reservoirs possess distinct characteristics, the fracture treatment implementation needs to be different and therefore meticulous strategy for well stimulation is needed. Fracture treatments in cased-hole completions are primarily required for following reasons:

- Invasion of the natural fractures for CSG/CBM production is necessary to unclod the drilling fluid and cement. Hydrofracturing removes such plugging and connects the wellbore to the natural fracture system.
- Pressure drawdown from dewatering must be extended deep into the coal seam to accelerate the release of gas from the micropores. Gas production can be accelerated by a propped fracture which shall extend the effective borehole radius.

- Hydrofracturing reduces the pressure in the near well-bore area and helps in distributing the pressure drawdown to reduce fines production.
- Hydrofracturing effectively connects the wellbore to the entire reservoir which may not be achieved through perforations, even at high density.

Hydraulic fracturing of a coal seam is mainly a process that interconnects the cleat system of the coal to the wellbore. For extremely low permeable reservoirs long and conductive fractures must be created to achieve economic flow rates. Bottom hole pressure during production should be minimized to accelerate desorption of gas. A wide, conductive propped fracture system that can withstand the maximum closure stress must be created to minimize the pressure drop down in the fracture. In hydraulic fracturing of coal seams, because of high treating pressures and complex nature of the fracture systems, real-time changes during the treatment may become the rule rather than the exception.

Multizone entry in cased holes can be achieved by baffled entry, frack plug entry, partings entry or coiled tubing, and packer completions technique. Baffle plates are often placed on the casing before installing and cementing the casing for fracturing individual locations in between coal seams. Several sequences of operations involving perforation and fracturing of the bottom seam at different shots per length intervals have been developed. Then baffle plated are dropped to seal off the bottom seam to aid perforation and stimulation of the target seam directly above. The procedure is continued until all the seams above the bottom seams are stimulated. Hollow frack plugs which allow the operator to set a plug-type tool on electric wireline in the casing after perforating a zone above are used. A ball, much like the one used in the baffle technique, could be dropped or placed in the top of the tool to allow shutoff of the casing below when initiation of the next frack stage above has commenced. A number of plugs can be set to allow for individual staging of each treatment.

A variety of fracturing fluids including crosslinked and uncrosslinked gelled water, commingled CO_2 gelled water, CO_2 foam and N_2 foam has been used though the borate crosslinked gelled water made ultra-frack III-35 is most widely used. The frack fluid should possess the following properties:

- Crosslinks rapidly yet is shear recoverable
- Has high viscosity during pumping time
- Has controllable gel break
- Incorporates additives that dramatically reduce filter-cake residue.

Proppants are mainly used to prevent pack plugging and loss of fracture conductivity due to coal fines. To maximize fracture conductivity and minimize the consequences of proppant embedment into the soft coal, high concentrations of proppant (i.e., 8–10 ppg) have frequently been used.

7.4.4 PRODUCTION STAGES

A typical CBM well production profile is illustrated in Figure 7.8. Unlike conventional reservoirs, a CBM reservoir needs to be dewatered first. The cleat space is initially occupied with water. As reservoir pressure depletes, the gas in coal matrix desorbs and flows into the cleat space.

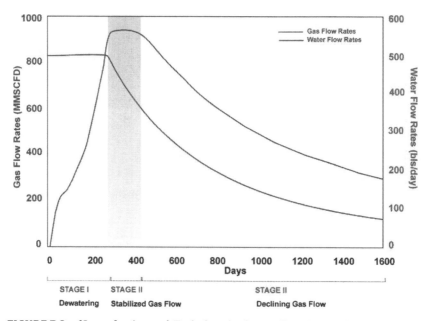

FIGURE 7.8   (See color insert.) Typical production profile and stages from a CBM well.

The production life of a CBM well can be divided into three stages (Fig. 7.8):

Stage 1: This is the initial stage of production life. This stage is also known as dewatering phase. In this phase of production life, the reservoir pressure is depleted below desorption pressure through dewatering the formation. The gas saturation increases in the reservoir causing the gas rate to increase and reach a peak.

Stage 2: This stage is stabilized gas production stage. The gas saturation develops in the specific areas far from the wellbore which supports sustained and stabilized production for some duration.

Stage 3: During this stage gas production declines after gas saturation approaches maximum in entire reservoir. The water rate decreases significantly and reaches minimal.

7.5 ENVIRONMENTAL ISSUES

7.5.1 CO-PRODUCED WATER, ITS TREATMENT, AND DISPOSAL

Normally, water must be removed from the coal to lower the pressure and to initiate methane desorption. However, near mining operations, there may be only small amounts of water to produce. The main environmental problem in CBM projects is the effect on water resources. Water is crucial for mankind without which the human race cannot survive. The physical and chemical properties of produced water vary considerably depending on the geographic location of the field, the geological host formation, and the type of hydrocarbon product being produced. Properties and volume of produced water can even vary over the life of a well. A large amount of water production is anticipated early in the life of a well thereby posing a financial burden on the operator in the first few years. Thereafter the water production gradually declines to its lowest level toward the end. Consequently, the water disposal issue reduces with time.

Major constituents of concern in CBM co-produced water is salt content (expressed as salinity, total dissolved solids, or electrical conductivity), various natural inorganic and organic compounds (e.g., chemicals that cause hardness and scaling such as calcium, magnesium, sulfates, and barium) and naturally occurring radioactive material (NORM). High levels of salinity can affect the vegetation's ability to withdraw water from the soil. The accepted level of salinity for irrigation is around 3 dS/m. In addition to the negative effect of sodicity on the soil structure, high levels of Na^+ may cause a reduction in Ca^{2+} and Mg^{2+} in the plants if used for irrigation. Irrigation with water of high Na^+ concentration may affect the soil properties such as soil infiltration, permeability, and aggregate stability. Thus, water contamination is a potential concern associated with the disposal of CBM co-produced water and requires prior treatment for its salinity and sodicity to meet water quality standards.

CBM co-produced water may affect the surface water quality, underground water, and the overall ecosystem's balance. It may reach water channels in several ways and degrade the surface water due to excessive salt content. CBM water percolates the unlined infiltration ponds into permeable surface sediments into the groundwater aquifers causing possible groundwater contamination. The produced water that is discharged directly into rivers and streams can also make its way into the groundwater in case of influent streams.

Selection of a management option for produced water at a particular site varies based on the following factors:

- Physicochemical properties of the co-produced water
- Flow parameters (rate, volume, and duration) of water produced
- Local aspiration of utilizing the co-produced water
- Permitted treatment and disposal options as specified by regulatory authority
- Technical and economic feasibility of suitable options
- Infrastructure availability of suitable disposal option.

Broadly, there are four options for handling CBM co-produced waters: (1) discharge into surface streams, (2) use in land applications, (3) re-injection into the wells, and (4) treatment through membrane processes. Some of the treatment options those are in practice in CBM industries are discussed below.

Reverse osmosis is used to remove dissolved salt and this utilizes cross filtration of high TDS water through a high pressure, semipermeable, cellophane-like membrane. The treated water usually meets safe drinking water standards and is of high quality with about 95–99% of dissolved salts, total organic carbon, and silica removed. Reverse osmosis pressurized filtration through the fine-pore membrane is highly effective, can remove most contaminants, minimizes waste stream for disposal and is a proven technology. The disadvantage of this technique is that a huge volume of untreated water is bypassed. Reverse osmosis involves high capital expense owing to the expensive membrane and installation of significant infrastructure. The membranes are prone to fouling and sensitive to temperature, they may require pretreatment. The technique involves relatively immobile infrastructure, high per-unit treatment as well as waste stream disposal cost.

Electrodialysis (ED) or electrodialysis reversal (EDR) are electrochemically driven desalination techniques which involve separation of salts through ion exchange membranes. A series of ion exchange membranes are used with electrically charged functional sites which are arranged in an alternating mode between the anode and the cathode which remove charged substances from the feed water. A positively charged membrane allows only anions are allowed to pass through it whereas negatively charged membranes allow only cations to pass through them. Periodic reversal of polarity is implemented to optimize the filtration operation. Both EDR and ED technologies have only been tested on a laboratory scale for the treatment of produced water and are found to work best for treating low saline produced water. With the membrane lifeline between 4 and 5 years, regular membrane fouling and high treatments costs limit this technology to a great extent.

Distillation process involves boiling water to form vapor which is then passed through a cooling chamber for condensation into water again. This produced 99.5% pure water and contaminants like bacteria, dissolved solids, heavy metals, nitrates, sodium, and many organics are removed. Distillation is effective as compared to reverse osmosis and other treatment options but is associated with high energy consumption for vaporizing the water.

Evaporation pond methods impound the co-produced water is an artificial pond and water is allowed to evaporate naturally by sun's heat. Ponds are designed to prevent subsurface infiltration and groundwater contamination. It is a favorable technology for warm and dry climates where evaporation occurs at accelerated rates. Evaporation ponds are typically economical and have been widely employed both onsite and offsite the CBM production area. The only limitation is that the co-produced water is not usable as it is lost through evaporation.

Ion exchange is a widely applied technology in industrial operations for various purposes, including utilization for the treatment of CBM the co-produced water. It is especially useful for the removal of monovalent and divalent ions and metals by resins from co-produced water. Ion exchange technology has a lifespan of ~8 years and will require pretreatment options for solid removal. It also requires the use of chemicals for resin regeneration and disinfection.

7.6   FUTURE OF COAL SEAM GAS

7.6.1   *GLOBAL RESOURCE AND RESERVE ESTIMATES*

CBM development was started in the early 1980s. In recent times, it has gained prominence in countries like the United States, Australia, China, Russia, and Canada. In 2006, World Coal Organization has estimated the global CBM resources total of 143 TCM of which only 1 TCM was actually recovered. By the end of the year 1990, coalbed methane reserve in the United States was measured about more than 700 TCF with 100 TCF of economically recoverable gas-in-place. Among the recoverable CBM reserve of the United States, 57 TCF CBM was estimated in Alaska. Gas-in-place was estimated in Russia was ranging from 600 to 2825 TCF. China has the world's third-largest ready reserves of CBM with an estimated reserve of 36.8 TCM among of which 10 TCM can be exploited. Southeast British Columbia has a proven reserve of CBM resource for more than 12 TCF and 8 TCF for northwest British Columbia. A resource of about 60 TCF was found in North East British Columbia and 0.3 to 1.6 TCF reserve was found in Vancouver Island.

Production of coalbed methane was first started in 1971 in the underground coal mines for the reduction of explosive gas hazards. The United States started exploratory CBM well in the year 1984. By the end of the year 2000, the United States explored a total number of 13,973 coalbed methane production wells and CBM production was about 7% of the total United States dry gas production and 9% of proven dry gas reserves. The United States is the global leader in CBM production and has produced over 1.91TCF in the year 2009. China government has set up a target production of 50 BCM by 2020. According to world coal organization report, China has produced 402 BCF CBM in 2011 whereas the production has increased to 542 BCF in the year 2012. Australian Petroleum Production & Exploration Association Limited has estimated an annual CBM production of 4 BCM for the year 2008 in Australia with increasing rate of 39% per year from 1990. In 2011, Australia's gross CBM production was 252 BCF. Commercial CBM production of India started in the year 2007 by

Great Eastern Energy Corporation Ltd and in 2011 it was 10.5 BCF while it was just over 2 BCF in the year 2009. The prognosticated CBM resource of India is about 92 TCF. As per the Director General of Hydrocarbons (DGH) India, the current CBM production is around 0.77 MMSCMD.

7.6.2 CLEAN ENERGY TECHNOLOGIES

7.6.2.1 COAL TO LIQUID

The high oil price since the early 1990s has prompted the latest efforts in coal to liquid (CTL) development worldwide and attracted attention to the consumers as well as investors. Coal is a solid fossil fuel having high carbon content and low hydrogen content of only about 5%. Whereas, typically used transportation fuel extracted from the crude oil having almost double hydrogen content. Hence, producing oil is not the only objective. Maintaining similar hydrogen contents to that of oil with similar properties is also a challenge. This can be achieved by removing carbon and/or by adding hydrogen, either directly or indirectly, while reducing the molecular size. Also during the process, elements such as sulfur, nitrogen, and oxygen present in coal must be largely eliminated. Thus, the technical challenge is to increase the hydrogen/carbon (H/C) ratio in the product. In the direct liquefaction process, coal is dissolved in a solvent at high temperature and pressure. This process is highly efficient, but the liquid products require further refining to achieve high-grade fuel characteristics. However, in the indirect method of liquefaction coal is gasified to form "syngas" (a mixture of hydrogen and carbon monoxide). The syngas is then condensed over a catalyst through the Fischer–Tropsch (FT) process to produce high quality, ultra-clean products. South Africa based company Sasol has established both the process in several countries.

7.6.2.2 UNDERGROUND COAL GASIFICATION

Gasification of coal for power generation is becoming one of the high priority research areas. Enhanced utilization of the domestic coal resource is necessary to reduce the dependence on import of fossil fuel for many countries while achieving reduced emissions of greenhouse gases. Underground coal gasification (UCG) technology can be used to produce clean energy using domestic coal resources, including currently un-mineable deposits, at competitive costs with minimum damage to the environment.

UCG is a thermochemical process of in situ gasification of coal aiming at the recovery of the heat value of coal for power generation and recovery of usable gaseous products as feedstocks for chemicals and fertilizers. UCG eliminates the need of mining for exploitation of coal as the products are recovered through the in situ gasification route. It is achieved through a set of injection and production wells (Fig. 7.9) by injecting oxidants, gasifying the coal in situ and bringing the product synthetic gas (syngas) to the surface. Syngas may be used as fuel and for a wide range of chemical syntheses. Depending upon the temperature and availability of O_2 the type of reaction varies and accordingly three dominant zones along the length between two wells can be identified depending on the dominant reaction. The simultaneous process of heating, drying, pyrolysis, and gasification takes place to generate syngas, combustible components of CO, H_2, CH_2, with other hydrocarbons like CO_2, H_2O, and N_2 which are less desirable products.

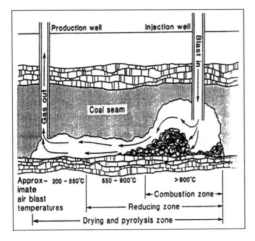

UCG Gasification reactions
Combustion of Carbon: $C+O_2 \rightarrow CO_2$
Partial Oxidation: $C+1/2O_2 \rightarrow CO$
Oxidation of CO: $2CO+O_2 \rightarrow CO_2$
Water gas shift reaction: $CO+H_2O \rightarrow CO_2+H_2$
Methanation reaction: $CO+3H_2 \rightarrow CH_4+H_2O$
Hydrogenation reaction: $C+2H_2 \rightarrow CH_4$
Boudouard reaction: $C+CO_2 \rightarrow 2CO$
Steam Carbon reaction: $C+H_2O \rightarrow H_2+CO$

FIGURE 7.9   Schematic representation of UCG process with set gasification reactions.

Environmental hazards associated with conventional underground mining/surface strip mining or surface gasifiers operations are eliminated by UCG. Coal with high ash content is effectively utilized by this method minimizing its environmental impact through its excavation and mining. At the same time, it does not require management of ash produced by thermal power plants. In situ use of coal greatly reduces the need to mine and transport coal. UCG also emits fewer greenhouse gases and is an effective means of mitigating global warming and climate change. UCG combined with carbon capture and sequestration (CCS) than by addition of CCS to a coal-fired power plant. Overall, the carbon footprint and capital expense associated with UCG much less compared to conventional coal mining and utilization.

KEYWORDS

- coal
- methane
- adsorption
- fracking

- production
- UCG
- CCS

REFERENCES

Burton, E.; Friedmann, J.; Upadhye, R. Best Practices in Underground Coal Gasification. Lawrence Livermore National Laboratory Contract No. W-7405-Eng-48, 2004, 119.

Chattaraj, S.; Mohanty, D.; Kumar, T.; Halder, G. Thermodynamics, Kinetics, and Modeling of Sorption Behavior of Coalbed Methane—A Review. *J. Unconven. Oil Gas Resour.* **2016,** *16*, 14–33.

Creedy, D. P.; Garner, K.; Holloway, S.; Jones, N.; Ren, T. X.; 2001. *Review of Underground Coal Gasification Technological Advancements*; Report No. COAL R211, DTI/Pub URN 01/1041, UK DTI Cleaner Coal Technology Transfer Program: London, UK, 2001.

Flores, R. M.; Rice, C. A.; Stricker, G. D.; Warden, A.; Ellis, M. S. Methanogenic Pathways of Coal-bed Gas in the Powder River Basin, United States: The Geologic Factor. *Int. J. Coal Geol.* **2008,** *76* (1), 52–75.

Ma Y.; Holditch S. *Unconventional Oil and Gas Resources Handbook*, 1st ed.; Evaluation and Development, Gulf Professional Publishing: Waltham, USA, 2015, p 550.

Mohanty D. Geologic and Genetic Aspects of Coal Seam Methane (Chapter 4). In *First Indo-US Workshop on Coal Mine Methane*, Singh A. K.; Mohanty D., Eds.; 17–20 October 2011, pp 45–60.

Mohanty D. In An Overview of the Geological Controls in Underground Coal Gasification. In: *UCG-2017, IOP Conf. Series: Earth and Environmental Science*, 76, 012010, 2017, p 8.

Rogers, R. E. *Coalbed Methane: Principle and Practices*; PTR Prentice Hall: Englewood Clifts, NJ, 07632, 1994, p 345.

Saulsberry, J. L.; Schafer, P. S.; Schraufnagel, R. A. *Coalbed Methane: Principles and Practices—A Guide to Coalbed Methane Reservoir Engineering*; Gas Research Institute: Chicago, Illinois 60631, USA, Halliburton, 1996.

Scott, A. R. Composition and origin of coalbed gases from selected basins in the United States, in Proceedings from the 1993 International Coalbed Methane Symposium, The University of Alabama, Tuscaloosa, May 17-21, 1993, p 207–222.

Seidle, J. *Fundamentals of Coalbed Methane Reservoir Engineering*; PanWell Books: Oklahoma, USA, 2011, p 401.

Thakur, P. *Advanced Reservoir and Production Engineering for Coal Bed Methane*; Gulf Professional Publishing: Cambridge, USA, 2016, p 208.

Whiticar M. J. A geochemial perspective of natural gas and atmospheric methane. Advances in Organic Geochemistry 1989. Org. Geochem. Vol. 16, Nos 1-3, p 531–547, 1990.

CHAPTER 8

Minerals as Sources of Energy

ABANI RANJAN SAMAL[*]

GeoGlobal, LLC, Salt Lake City, Utah, USA

[*]*E-mail: arsamal@gmail.com*

ABSTRACT

With increasing demand for energy, the mix of energy resources is evolving with time. Minerals, namely coal and uranium, has been a part of this mix for a long time. This chapter introduces these two minerals, discussing their geology, process of mineralization, and their exploration and mining. These two minerals are very different to each other in terms of their underlying geology. The gradual process of coalification results in different grades of coal which give rise to a coal-ranking system. Mineralization of uranium-bearing minerals and its geological assessment is very different and so is its mining. Both coal and uranium-bearing minerals are mined but uranium can also be extracted using in situ leaching. Supply and demand of these commodities, their exploration and extraction methods are included in this chapter. A discussion of exploration techniques and their reserve assessment is also included. The chapter concludes with a discussion on guidelines on reporting exploration results, mineral resources and reserves, and at the end provides references to support the interested reader to learn more.

8.1 INTRODUCTION

Minerals play a big role in meeting the ever-increasing demand for energy. There are two major minerals which are used to produce energy. We call them "energy mineral" in this chapter. These minerals provide used for heating homes and offices and in the manufacture of various products that we use and provide fuel for transportation. The solid energy minerals resources include coal and uranium. The coal is a readily combustible rock. The uranium is a radioactive element that occurs in a mineral form. The important uranium-bearing mineral is pitchblende.

For comparison purposes, Table 8.1 shows the amount of power that can be derived from various sources such as firewood, coal, and uranium. According to the Nuclear Energy Institute (NEI, https://www.nei.org), a single uranium fuel pellet, which is of the size of a fingertip contains as much energy as 17,000 cubic feet of natural gas, 1780 pounds of coal or 149 gallons of oil.

TABLE 8.1   A Comparison of the Amount of Power that Can Be Generated from Various Sources.

Source	Estimated amount of power
1 kg of firewood	1 kWh power
1 kg of coal	3 kWh power
1 kg of oil	4 kWh power
1 kg of uranium	50,000 kWh power

8.2   SUPPLY AND DEMAND

Coal as a source of energy still provides a major role in the mix of various sources of energy. The global demand is set to increase by 15% by 2040. Majority of the coal is produced in China, India, Indonesia, and Australia, which may increase up to 70% of the world's coal production by 2040. China being the largest consumer of coal may consume 50% or more of the global production of coal, which may decline after 2030 (IEA, 2014). India is the second largest coal consumer. India's demand for coal has been increasing over last decade. The United States of America, Russia, China, Australia, and India area the top five countries in terms of coal reserves.

The International Energy Agency (IEA) predicts that in next two decades the demand for more nuclear power is projected to increase by economies such as China, India, Korea, Russia, and the United States. Uranium is the only fuel used in the nuclear fuel sector and is responsible for approximately 12% of world's energy supply (Minerals Make Life, 2016). The major producers of uranium are Australia, Kazakhstan, Canada, Russian Federation, South Africa, and Namibia (Figs. 8.1–8.5).

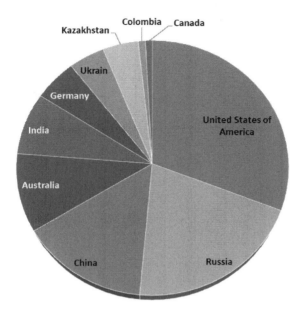

FIGURE 8.1 **(See color insert.)** Major coal producers in the world.

Source: http://www.mining-technology.com/features/feature-the-worlds-biggest-coal-reserves-by-country/

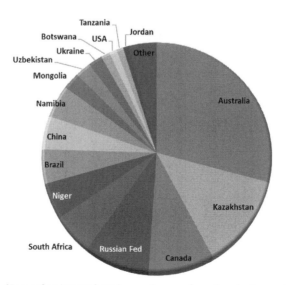

FIGURE 8.2 **(See color insert.)** Major producers of uranium in the world.

Source: http://www.world-nuclear.org/information-library/nuclear-fuel-cycle/uranium-resources/supply-of-uranium.aspx

FIGURE 8.3   A geologist collecting data in an exploration project in the western USA.

FIGURE 8.4   A trial pit in an exploration project (left). A core-drilling machine at work at a coal deposit (right).

The coal and uranium deposits are developed to meet the demand of energy from these sources. In the following sections, we will discuss the general depositional environments, resource assessments, and extraction procedures of coal and uranium.

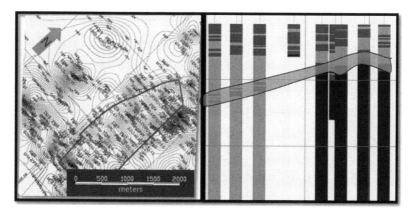

FIGURE 8.5 An example of a topographic map with the location of drill-holes and potentials for coal identified as closed red polygons (left). A vertical cross-section of drill holes showing coal intercepts captured by a shaded polygon (right).

8.3 MINERAL EXPLORATION: GENERAL OVERVIEW

This section of the will provide a general summary of mineral exploration, which then will be referenced in the discussion of exploration for coal and uranium mineralization at later sections.

Mineral exploration is a multifaceted endeavor that involves several phases of tasks. Broadly there are four phases:

1. *Permitting and pre-field logistical coordination:* This is the phase that ensures that the access to the land and all tools and accessories for the field mapping is also available. The access to the land means that the exploration agency has legal rights to the land granted through and by the appropriate authority at local, state and federal levels for carrying out necessary investigations.

2. *Field exploration:* This phase consists of many tasks such as mapping, sampling analysis, and compiling results. The mapping the local and regional geology is one of the foremost important and time-consuming tasks. A geologist spends a substantial amount of time in collecting data and samples which are later used for creating mineral prospect maps. Fundamental knowledge of geology and mapping are required to be able to discover the deposit and assess the potential of the deposit for next stage. Various type of technique is applied:

   • Geology: Rock types and structures

- Geochemical
- Geophysical
- Trenching and pitting
- Drilling
- Underground shaft and drifts

3. *Geological interpretation:* Based on the data and samples collected from field investigations, detailed geological maps are created with areas of mineral potentials identified. Cross sections and 3D models are created for use at a later stage.

4. *Resource estimation and life of mine design*: At this stage, the data generated from the exploration campaigns are used for assessment of the total amount of resources that have *an eventual prospect for economic extraction*. Only the amount of material meeting this criterion can be subject to a life of mine planning. A life of mine (LoM) plan is a plan for extraction of material from a mining project at a profit over a certain time period. The 'life' of a mine is estimated the number of years that is required to operate the mining project. The LoM may be updated with new information such as cost or production, the price of the product, legal issue or social issues. For an example, if a mine was scheduled to be mined for next 20 years, but a lowering of the price of the commodity may force a mining project to close its operation much prior to the initial planning date. Generally, the cost of production of one ton of rock is higher in an underground mining operation (Fig. 8.6) compared to a surface (or open cast) mining operation (Fig. 8.7).

This discussion of mineral exploration is common to coal and uranium. However, in the coal exploration sections, specific topics relevant to those commodities are highlighted in the following sections.

8.3.1   *GEOLOGY OF COAL DEPOSITS*

The coal forms by compaction and alteration of plant remains. Coal is known to form under non-marine conditions. The process of formation of coal is called coalification. The coalification goes through two transitional stages known as the biochemical and physicochemical stage of coalification (Diessel, 1992). The biochemical coalification process is the initial stage of the process in which the plant remains to go through the chemical

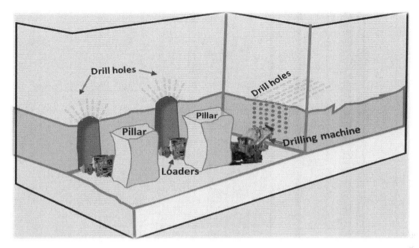

FIGURE 8.6 A schematic diagram of an underground mine. This figure is a simplistic representation of a room and pillar-type mining method.

FIGURE 8.7 A photograph of an open cast mine or surface mine. The above picture shows layers of coal being extracted.

decomposition and converted into peat and brown coal (Fig. 8.8). In the subsequent physicochemical transformation stage, the peat formed in the earlier stage is transformed into various coal types. This second stage is also known as the geochemical stage (Jones and Godfroy, 2002). The biochemical process is also known as the diagenetic stage.

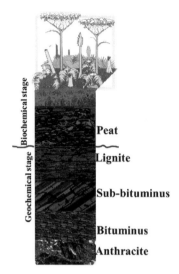

FIGURE 8.8   The coalification process.

The peat formed at the biogeochemical stage is decayed plant material which over time becomes soil-like material (texturally). This is a partial decomposition of plant material which is also known as "peatification."

The peat goes under continued burial and exposes the peat material to high temperature and pressure. The peat is compacted and loses water. In the process the coal forms with the increased richness of carbon in the material. As the coal is enriched in carbon various products like lignite, sub-bituminous, and bituminous coal and anthracite coal are formed. The process of coalification is irreversible.

Coal classification-based rank: Coal is classified into various categories or ranks based on quality. Coal in the increasing order of rank: lignite, subbituminous, bituminous, and anthracite coal. Figure 8.9 shows the ranking system used in the United States. The lignite is the lowest rank coal and anthracite with high calorific value, lowest volatile content.

8.3.2   COAL EXPLORATION AND RESOURCE ASSESSMENT

Mapping for coal exploration focuses on measurements of coal bed thicknesses and lateral extent, orientations of the coal beds, contacts between different rock units, and structures (faults, fractures, etc.). The coal outcrops are well documented.

			Calorific Value	Volatile matter
High Rank	Anthracite	Meta-Anthracite		Nearly 0
		Anthracite		
		Semi-Anthracite	> 14,000	*Decreases*
Medium Rank	Bituminous	Low-Volatile		
		Medium Volatile		31
		High Volatile A	14,000	
		High Volatile B		
		High Volatile C		
Low Rank	Sub-Bituminous	A	*Increases*	
		B		
		C		
	Lignite	B		
		A	5,000	
	Peat			

FIGURE 8.9   The coal ranking system in the United States.

The drilling and geophysical surveys for the coal resources are based on the results of the mapping campaign. The geophysical methods include surface geophysical mappings such as air-borne, magnetometer investigations, regional gravity surveys and broad-scale seismic studies which are used to delineate the sedimentary and structural framework in a regional scale and provide guidance for target generation. The gravity surveys are instrumental in providing major sedimentary units of the depositional sedimentary basin of the coal deposit. The magnetic survey depends on the existence of contrasting magnetic properties between different rock layers. The airborne magnetic surveys are used for regional scale mapping, whereas the detailed magnetic surveys are useful for closer investigations of the structural units such as joints and fractures that are of local scale interest. The Electrical resistivity is dependent on the resistance of the rock units to electricity, which partly depends on the properties of the constituent minerals and/or, the presence of fluid in the pore spaces of the rock unit. The electromagnetic methods are applied to map the depth of the coal depositing basement, to identify the coal-bearing rock units or individual coal seams. These are highly effective but expensive techniques. The seismic techniques provide very important information in coal exploration: delineating the broad structural features of a relatively large area and mapping out individual small-scale structures, such as faults, splits, and washouts.

The drilling can be noncore or core drilling. The noncore drilling data are used in geological logging. The core drilling or diamond drilling is more expensive but highly effective in providing key information such as lithology, structures, recovery, hardness, etc. The pictures in Figure 8.10 show an example of good and poor recovery of coal in a diamond drilling campaign for lignite.

Both geological and geophysical data are used to guide subsequent exploration drilling programs. These data are also used in detailed 3D modeling and resource estimation. The geological model is created by correlating different coal-bearing and waste material (usually sandstone or shale) between drill hole data, similar to what is shown in Figure 8.11. Using various software tools, three dimensional (3D) geological model is constructed which allows visualizing the geological horizons on a computer screen. This geological model is used in the assessment of mineral resources in a coal deposit.

Estimation of grades such as carbon content, ash content, moisture, sulfur, and so on are done using interpolation of drill hole data. Amongst various interpolation techniques used, the inverse distance square (IDS) and kriging technique are very popular.

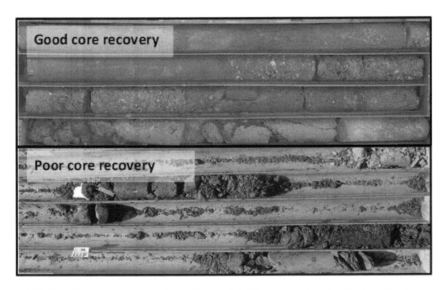

FIGURE 8.10 Core recovery in a diamond drilling campaign for lignite. Good core recovery due to competent rock types (top). The coal and finer and lighter sandy units are missing (bottom).

FIGURE 8.11 **(See color insert.)** A schematic presentation of a geological model of a coal deposit.

In the inverse distance power interpolation technique, the weights on the samples used for any estimation is inversely proportional to the distance of separation as explained in Figure 8.12. In these techniques, when the power equals to 2, it is known as IDS, when power equals to 3, it is inverse distance cubed, and so on. An example of the application of IDS is provided in Figure 8.13. The IDS technique is known to have "bulls eye" effects at locations of high and low-grade values.

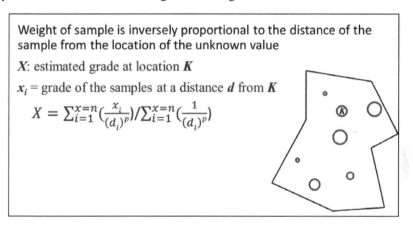

Weight of sample is inversely proportional to the distance of the sample from the location of the unknown value

X: estimated grade at location K

x_i = grade of the samples at a distance d from K

$$X = \sum_{i=1}^{x=n}\left(\frac{x_i}{(d_i)^p}\right) / \sum_{i=1}^{x=n}\left(\frac{1}{(d_i)^p}\right)$$

FIGURE 8.12 Explanation of inverse distance square technique.

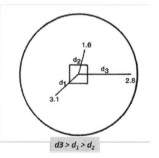

% Weighting to each Grade				Estimated grade
	d_2	d_1	d_3	
p	1.60%	3.10%	2.80%	
1	44.8	34.7	20.5	2.37
2	55.2	33.2	11.6	2.34
3	64	29.9	6.1	2.12

FIGURE 8.13 An example of application of the IDS technique (Annels, 1991).

Kriging is a geostatistical interpolation technique which is popularly known for its simplicity (Linear) and unbiasedness that is, estimates are independent of the actual value of the sample points (input data), rather dependent on the spatial arrangements of these input sample points. Kriging algorithm ensures minimum variance and minimum error of estimates. The weights on the samples (*w*) depend on the spatial model of the dataset. There are various types of kriging used in mineral industry. They are ordinary kriging (OK), simple kriging (SK), indicator kriging (IM), and multiple indicator kriging (MIK). The OK is the most popular, as it is the simplest technique and provides unbiased estimates. The kriging technique is named after famous South African Geostatistician Dr. D G Krig. According to Krig, "geology should form the foundation of any geostatistical analysis" (Krige, 1999). The kriging technique, being a geostatistical technique, should be applied only if the underlying assumptions of second-order stationarity are met, that is, at a minimum the mean and variance of the sample data remain invariant in space. If the mean and variance of the samples do not meet these stationarity assumptions, it is advisable to apply an interpolation technique other than kriging.

In a comparative study (Hannon and Sherwood, 1986), it was found "the geostatistical method that is, kriging gave the best estimate of the overall tonnage. Both IDS. and the kriging methods gave a reasonable estimate of mineable coal in selected blocks." The readers are also encouraged to refer to the Ph.D. thesis of Knudsen (1945).

8.3.3   GEOLOGY OF URANIUM DEPOSITS

The International Atomic Energy Agency (IAEA) proposed two types of uranium deposits: *Conventional* uranium resources are those from which uranium is recoverable as a primary product, a co-product or an important by-product. *Unconventional* resources are resources from which uranium is only recoverable as a minor byproduct, such as uranium associated with phosphate rocks, nonferrous ores, carbonatite, black shale and lignite (Red Book 2012, IAEA).

The uranium occurs in oxidizing environments. The uranium-bearing minerals are preferable of layered structures as the $A(UO_2)RO_4xH_2O$ or $B(UO_2)CO_3xH_2O$ units, where A and B can be Na, K, Ba, Mg, Cu, Fe, Fe^{2+}, or Pb and $R = P$, As or V.

- Common uranium minerals are:
- Carnotite: $K_2(UO_2)_2(VO_4)_2 \cdot 3H_2O$
- Tyuyamunite: $Ca(UO_2)_2(V_2O_4)_2 \cdot 8 H_2O$.
- Autunite: $Ca(UO_2)_2(PO_4)_2 \cdot 10H_2O$
- Torbernite: $Cu(UO_2)_2(PO_4)_2 \cdot 12 H_2O$
- Uranophane: $Ca(UO_2)_2SiO_3(OH)_2 \cdot 5(H_2O)$
- Zeunerite: $Cu(UO_2)_2(AsO_4)_2 \cdot 12(H_2O)$

Sandstone-hosted uranium deposits: the sandstone-hosted uranium deposits are well known in the United States, Canada, and Europe as listed in Table 8.3.

A schematic diagram of the geology of a sandstone-hosted uranium deposit is shown in Figure 8.14. The uranium occurrence is lithology controlled. A description of sandstone-hosted mineralization is also provided by Fewster (2009).

TABLE 8.2   The IAEA Classification of the Uranium Deposits.

	Conventional resources	**Unconventional resources**
1	Intrusive	Intrusive anatectic and plutonic
2	Granite-related	Granite-related
3	Polymetallic hematite breccia complex	Polymetallic hematite breccia complex
4	Volcanic-related	Volcanic-related
5	Metasomatite	Metasomatite
6	Metamorphite	Metamorphite
7	Proterozoic unconformity	Proterozoic unconformity
8	Collapse breccia pipe	Collapse breccia pipe
9	Sandstone	Sandstone
10	Paleo-quartz pebble conglomerate	Paleo-quartz pebble conglomerate
11	Surficial	Surficial
12	Coal-lignite	Coal-lignite
13	Carbonate	Carbonate
14	Phosphate	Phosphate
15	Black shales	Black shale
16	Phosphate	
17	Black shales	

Source: Adapted from ref. Bruneton (2014).

TABLE 8.3   Sandstone-Hosted Uranium Deposits in the United States, Canada, and Europe.

Powder River Basin

Shirley Basin

Gas Hills

Crooks Gap

Northern Black Hills District

Southern Black Hills District

Copper Mountain District (North Flank, Wind River Basin)

Poison Basin District (Southeast Flank, Washakie Basin)

Prior Mountain District (East Flank, Bighorn Basin)

Northwest-Southwest Williston Basin, South Dakota

Uravan Mineral Belt, Colorado Mineral Belt, Colorado-Utah Utah

Monument Valley-White Canyon District, Utah

Lisbon Valley, Utah

Grants-Laguna Districts, Zuni Uplift, New Mexico

Proterozoic rocks of East Great Slave Lake, Northwest Territories, Canada

Lower Permian Sedimentary Basins

Northern or sub Variscan Province, Northern Europe

Central or Variscan Province, Southwestern England, France, Central Germany, Czech Republic, Poland

Southern or Verrucano Province, Alpine and Apennine.

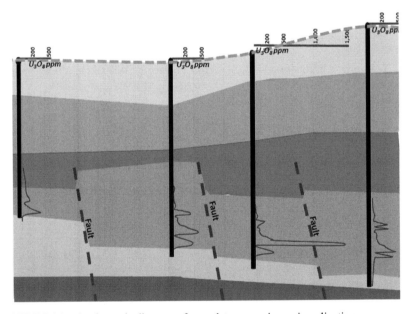

FIGURE 8.14   A schematic diagram of a sandstone uranium mineralization.

Roll front type sandstone-hosted uranium deposits: In certain sandstone-hosted deposits, uranium mineralization occurring below the water table is mobile in the direction of the water flow. These deposits form near regional oxidation/reduction fronts hosted in favorable organic-rich, porous sandstones below the water table. As shown in Figure 8.15, the uranium is enriched at the tip of the redox front.

8.3.4 URANIUM EXPLORATION AND RESOURCE ASSESSMENT

The exploration of uranium involves various geological, geochemical and geophysical techniques and relies on the geochemical and radioactive characteristics of the uranium mineralization. The geological techniques involve field mapping of the geological features such as rock-types, geological structures, and surficial exposures of the mineralization. The water, soil, and rock samples are collected and data are plotted on the geological maps. The geophysical surveys and prospecting are done from an aircraft or on the ground or in boreholes. The radiometric methods are highly effective in finding radiometric halos and are based on the radioactive properties of the uranium minerals. The geophysical methods used for deep subsurface exploration generally follow the surficial exploration methods. The process of measuring radiation is achieved through its ionizing properties. The instruments or sensors used in radiometric exploration are capable of converting radiometric radiation to an electrical signal, which is then used by the instrument for measuring the strength of the energy.

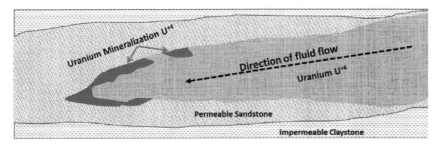

FIGURE 8.15   A schematic diagram of a roll-front uranium deposit.

A popular instrument used in uranium exploration is a Geiger–Müller (GM) counter (Fig. 8.16), which detects gamma rays; can be used on the surface and for borehole logging. A Geiger–Müller counter provides reading ion counts/per second. Higher intensity of uranium provides a higher count.

FIGURE 8.16   A Geiger–Muller counter.

The airborne techniques are highly useful in delineating regional anomalies (Wilford and Minty, 2009). For an example, the Radiometric Map of Australia was generated using airborne gamma-ray spectrometry over four decades which provides key information for uranium and thorium elemental enrichments at the regional scale. These maps are used for further detailed exploration at the local scale. Details of the airborne and ground geophysical exploration methods are well documented by the IAEA (2013).

Uranium resource estimation: The geology and style of mineralization of the uranium deposits are key factors for consideration in mineral resource estimation. The uranium deposits are geographically widespread and restricted to only favorable lithology and stratigraphic units. Understanding of the host rocks and their characteristics is a key. The ore may be preferentially replacing fossils and plant remnants. The mineralization may be overprinted by mild geochemical alteration and may be subparallel to the host sandstone layers. Structural displacements and structural controls of mineralization need to be fully understood so that the effects of the structural elements are implemented in the resource assessment.

Based on the complex nature of the mineralization, deterministic or probabilistic grade estimation techniques can be adopted. The polygonal method, the IDS, and kriging are deterministic techniques, whereas conditional simulation and indicator simulations/estimations can provide probabilistic grade estimates. The uranium mineral resources are assessed for mining using open-pit (or surface) or, underground mining methods (Fig. 8.17). As compared to coal deposits, the uranium mining requires hard-rock mining capabilities.

Where ore bodies lie in groundwater in the porous unconsolidated material (as shown in Fig. 8.20), the uranium can be extracted by dissolving the uranium underground (in situ) and pumping it out. This process is known as in situ leaches (ISL) mining or, (also known in North America as in situ recovery—ISR).

The ISL/ISR method is applied to uranium deposits where the mineralization occurs in a porous medium and is confined vertically and ideally horizontally by impermeable rocks. In Figure 8.18, the solid arrows show the injected fluid that brings the uranium into solution. The uranium-impregnated fluid is now pumped using a production well. Generally, the production well is surrounded by four to six injection wells.

FIGURE 8.17   An opencast uranium mine in Namibia.

The well-known uranium surface mining projects are Ranger in north Australia, Rössing in Namibia, and many uranium mines in Canada. The Olympic Dam in Australia, McArthur River, Rabbit Lake and Cigar Lake in Canada are underground mines. Mineral resource assessment considers the above-discussed mining methods and associated processing and extraction processes for assessment of criteria for eventual economic assessment.

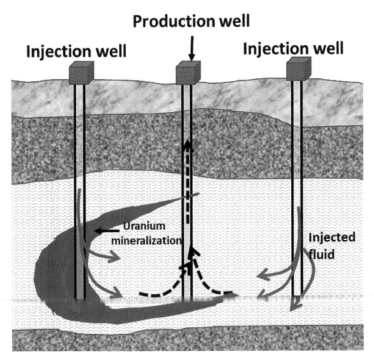

FIGURE 8.18 A generalized cross-section of a portion of a typical in situ leaching/ recovery operation.

8.4 FURTHER STUDIES IN COAL AND URANIUM RESOURCE ASSESSMENT

The "Australian Guidelines for the estimation and classification of Coal Resources 2014 Edition" is one of the leading documents available for practitioners for estimation and reporting of coal resources. This document provides a detailed guideline for exploration, estimation of resources, managing mineral inventory and reporting.

In addition to the mining method, extraction of the metal and handling of the radioactive waste material are also considered seriously while conducting various studies on a uranium deposit. The IAEA and other international authorities including ICMM (International Council on Mining and Metals) provide various guidelines for safe development, extraction, and use of uranium to meet the demand for energy.

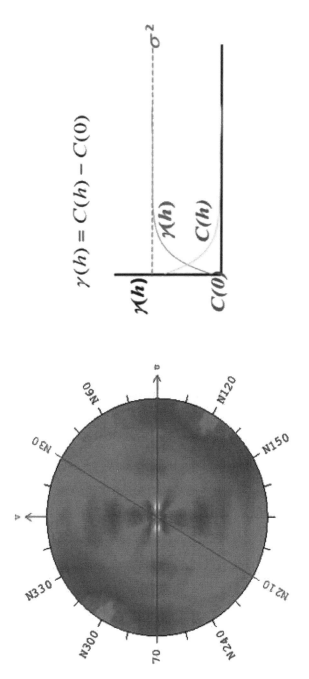

FIGURE 8.19 **(See color insert.)** A variogram map (left) and a variogram model (right).

8.4.1   GEOSTATISTICAL ESTIMATION PROCESS

1. Ensure that the data meets the second-order stationarity criteria

    *   Mean and variance remain unchanged in space
    *   The autocorrelation function depends solely on the degree of separation of the observations in time or space.

2. Develop a spatial continuity/structure model (such as variogram)
3. Use the variogram model (Fig. 8.19) in the estimation process such as OK or, SK or, geostatistical conditional simulations (CS)
4. Post process the models—in case of MIK, simulations, etc., nonlinear techniques.

The OK is the most popular geostatistical estimation technique. It does not require the parameters such as a "mean" (as it is required in SK). The OK is also known for being the best linear unbiased estimator or BLUE. As shown below, the OK system involves calculation of weights (w_i) and manage the errors in such a way that the error of estimation is minimum and the sum of all weights assigned to all samples used for estimation of an unknown value is 1.

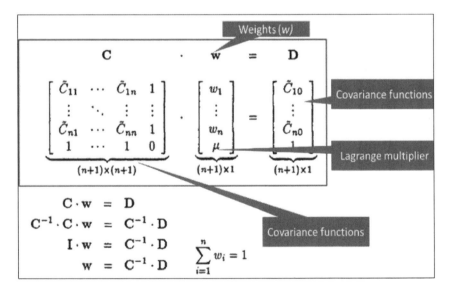

FIGURE 8.20   The Ordinary Kriging system involving matrix inversion for *n* samples used in the estimation of an unknown value.

KEYWORDS

- energy minerals
- coal
- uranium
- exploration
- mining

REFERENCES

Bruneton, P. (2014), IAEA Geological Classification of Uranium Deposits. International Symposium on Uranium Raw Material for the Nuclear Fuel Cycle: Exploration, Mining, Production, Supply and Demand, Economics and Environmental Issues, Vienna, Austria, 23 -27 June 2014 Conference ID: 46085 (CN-216).

Diessel, C. F. K. *The Coalification Process*. In: Coal-Bearing Depositional Systems; Springer: Berlin, Heidelberg, 1992.

Fewster, M. In *Sandstone Uranium Mineralisation associated with the Mulga Rock Deposits*. AusIMM International Uranium Conference 2009, Darwin, Northern Territory, 10–11 June 2009.

Hannon, P. J.; Sherwood, H. G. In *A Comparative Study of Geostatistical and Conventional Methods of Estimating Reserves and Quality in a Thin Coal Steam*. Proceedings of the Symposiums Sponsored by The Geology Division of ClM, and held in Montreal, Quebec, David, M.; Froidevaux, R.; Sinclair, A. J.; Vallee, M., Eds. May 10–11, 1986.

IEA's World Energy Outlook, 2014. Available at http://instituteforenergyresearch.org/analysis/ieas-world-energy-outlook-2014/ (accessed January 15, 2018).

Jones, J. C.; Godefroy. J. Stages in the Coalification Sequence Reflected in Oxidation Reactivities. *Int. J. Eng. Perfor. Based Fire Codes* **2002,** *4* (1), 10–12.

Knudsen, H. P. A Comparison of the Geostatistical Ore Reserve Estimation Method Over the Conventional Methods, 1945. Available at http://hdl.handle.net/10150/554906

Krige, D. G. Essential Basic Concepts in Mining Geostatistics and their Links with Geology and Classical Statistics *S. Afr. J. Geol.* **1999,** *102* (2), 147–151.

Minerals Help Us Meet Our Energy Needs, 2016. Available at http://mineralsmakelife.org/blog/minerals-help-us-meet-our-energy-needs/

Wilford J.; Minty B. The Use of Airborne Gamma-ray Imagery for Mapping Soils and Understanding Landscape Processes. In *Developments in Soil Science—Volume 31: Digital Soil Mapping – An Introductory Perspective*; Lagacherie, P.; McBratney, A. B.; Voltz, M., Eds.; Elsevier, 2007.

XiIAEA *Advances in Airborne and Ground Geophysical Methods for Uranium Exploration*, IAEA Nuclear Energy Series No. NF-T-1.5, 2013.

CHAPTER 9

Harvesting Energy from Wind

SANDEEP NARAYAN KUNDU[1,*] and MUNUKUTLA SRUTI KEERTI[2]

[1]*Department of Civil & Environmental Engineering,
National University of Singapore, Singapore*

[2]*SoUL Program, Indian Institute of Technology Bombay, Mumbai, India*

Corresponding author. E-mail: snkundu@gmail.com

ABSTRACT

Wind is a valuable natural resource which is harvested for generating electricity. Wind energy is a clean, renewable form of energy which supports energy production in rural low-income areas and is akin to a drought-resistant cash crop that has complemented rural income on top of agricultural practices by farmers and ranchers. Countries like the United States and Germany have successfully made use of wind power to complement the electricity production. However, wind power is intermittent and has unequal global distribution. This has given rise to different designs for wind turbines to suit different geographies in both urban and rural settings. The following chapter describes the fundamentals of wind, its distribution and analysis, power generation capacity, and its impacts and implications.

9.1 WIND POWER

Wind power is the ability to make electricity using the air flows that occur naturally in the earth's atmosphere. Wind turbine blades capture kinetic energy from the wind and turn it into mechanical energy, spinning a generator that creates electricity. Being a renewable resource, it is intermittent and its harvesting is done at different scale depending on the availability and the speed of wind. Utility-scale wind turbines are larger than 100 kW with connectivity to power grids and distribution systems for consumers. Distributed systems make use of turbines of 100 kW or smaller which are attached to the consumer and is meant for primary use. Offshore wind turbines erected in large bodies of water, usually on the continental shelf and have a huge potential for growth.

Wind energy is a clean, renewable form of energy that has many plus points. Wind power supports energy production in rural low income areas where most of the wind farms are located. Harvested energy from wind is akin to a drought-resistant cash crop that has complemented rural income on top of agricultural practices by farmers and ranchers. In the United States, farm owners make use of their unproductive land, making at least $245 million in lease payments when the let wind projects run on it. Increased wind energy production can decrease fossil fuel and result in meeting CO_2 emissions targets set by economies to curb global warming and climate change.

9.2   HISTORY OF WIND ENERGY

Wind has been used since ancient times for winnowing—to separate wheat and rice from husk—which is still practices in many parts of the world today (Swift-Hook, 2012). Rice and wheat paddy are threshed to detach the undesired husk from the grains and the mixture is dropped from a height. The denser food grains drop vertically whereas the husk is blown further away by the wind. Winnowing by wind has been mentioned in the Bible where it is traditionally ascribed to King David (1000–965 B.C.) which evidences that the practice was prevalent at least 3000 years ago. Sails mounted on masts have propelled boats along the Nile River as early as 5000 B.C. By 200 B.C., use of windmills has been reported in ancient Chinese society for scooping water from wells and streams. The use of vertical-axis windmills with sails for grinding grain in Persia and the Middle East have been mentioned in ancient books by the 11th century where they were used extensively for food production. In Europe, the Dutch led in using wind power for draining lakes and marshes in the Rhine River Delta. In the late 19th century, later settlers extended the use of windmills for pumping water, and eventually to generate electricity for homes and industry. American colonists used windmills to grind food grains and to pump water and even cut wood in saw mills. Postinvention of electricity, wind power was applied in modern settings where power could be generated remotely to light homes.

The largest known turbine, in the 1940s during World War II, was a 1.25-MW turbine located atop a Vermont hill and generated power for the locality. Wind electric turbines were prevalent in Denmark until late 1950s and were later abandoned due to the availability of cheaper alternatives to power like fossil oil.

Post 1970, when oil production peaked and started to decline in the United States, the search for alternative energy intensified and resulted in the reentry of wind-powered turbines into the power generation market. This period witnessed the US government's cooperation with industry to advance the technology and the commissioning of large commercial wind turbines. However, when global oil production surged in the 1980s and early 1990s, oil was cheap again and power generation from wind became uneconomical. Tax incentives and subsidies were introduced to keep it going as fossil fuel usage and environmental concerns forced economies to lower their carbon footprint.

In the present day, we have a wide range of wind-powered generators which operate in almost every size range, from small turbines to large, near-gigawatt-size offshore wind farms (Chitnis, 2014). Small turbines have opened a window for the small-scale wind farmer living in high rises which generated power enough to complement other dedicated sources like solar. At a larger scale, gigawatt size offshore farms produce commercial electricity to national electric transmission systems wind speed locale for power generation. Wind is the cleanest form of renewable energy with the littlest of environmental impacts.

9.3 GLOBAL ENERGY FOOTPRINT

Electricity is the final energy consumed by various anthropogenic activities like domestic cooking and heating, industrial machinery and transportation. Various fossil and non-fossil energy resources are used to produce electricity. Renewable resources accounted for a mere 19% of the total resources which includes fossil fuels and nuclear power (REN21, 2017).

9.3.1 RESOURCES IN ENERGY FOOTPRINT

The prime source of energy today is fossil fuels which account for over 78% of the energy footprint (Fig. 9.1). This is followed by renewables

(~19%) and the remainder is from nuclear power (2.3%). Of the renewables, traditional ones account for about 9% and the remaining are modern renewables like biomass and waster, geothermal, hydropower and solar, and wind. Wind energy, today accounts for only 0.2–0.3% of the global energy production.

9.3.2 GEOGRAPHICAL TRENDS AND LEADING PRODUCERS

The wind energy sector is growing exceptionally and the total generation has grown seven times in the last decade (Fig. 9.2). Production in Asia Pacific has surpassed to that of Europe and North America. A huge potential exists in Latin Americas, Africa, and Middle East. However, more than 80% of the produced wind energy is concentrated in just eight countries of the world with China in the lead with over 33% share of the total. Second to China's wind energy production of 145,000 MW is the United States of America which has a share of 17% with around 75,000 MW. The other countries in order of their decreasing share are Germany (~10%), India (~6%), Spain (5%), United Kingdom (3%), Canada (2.6%), and France (2.4%). High potential regions in Africa and the Middle East may soon start contributing to wind energy footprint in the coming years.

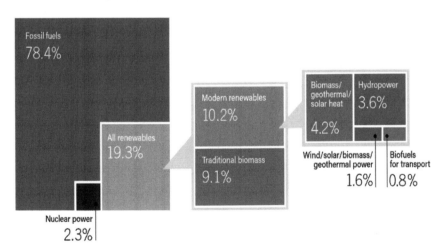

FIGURE 9.1 Estimated renewable energy share of global final energy consumption 2016. Wind power account for under 1% of global power generation.

Source: Reprinted with permission from REN21—Renewable 2017 Global Status report.

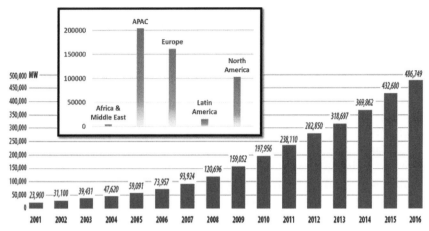

FIGURE 9.2 Annual global wind power capacity since 2001 and (inset) capacity by geographical region.

Source: Global Wind Energy Council 2016.

9.4 ORIGIN AND NATURE OF WIND

Meteorologically, winds are movements of air masses in our atmosphere. Winds operate on a very large scale and are primarily generated by atmospheric pressure differences due to differential solar heating.

9.4.1 PRESSURE, TEMPERATURE, AND VOLUME

Matter in the gaseous state can be compressed and also expanded by multiple times. Compression and expansion are effected by pressure which is defined as force per unit area. Earth's gravitational pull acts on gases in our atmosphere to create a force called atmospheric pressure.

The ideal gas law defines the relationship between pressure, temperature and volume. It is formulated as:

$$PV = nRT,$$

where P is pressure, V is volume, n is the number of moles, R is universal gas constant, and T is temperature. Differential solar insolation on Earth's surface is the prime driver of unequal temperatures over Earth's surface. As temperature is related to volume, hot surfaces result in expanded air masses and colder regions have denser air masses. Temperature differences influence pressure gradients which complements the Coriolis force which is associated with the rotation of the Earth, inducing mass movements of air in the atmosphere (Nfaoui, 2012). This results in wind systems which vary temporally, for example, daily, seasonally, and at higher decadal scales.

9.4.2   WIND VARIABILITY

Wind is not constant at any location. It varies in its speed and direction which depends on the geographical latitude and the distribution of land and ocean masses on Earth. The variation pattern of wind speed during 24 h of the day is called the diurnal cycle. Wind speeds, near the Earth's surface, are usually greater during the middle of the day than at night. This is due to solar heating, which warms the air making it rise higher up into the atmosphere. The emptied space is quickly replaced by cooler air from the sides resulting in thermal mixing. This causes wind speeds to slightly increase with height for the first few meters above the earth. The mixing stops at night time and the air near the surface comes to a stop, and the winds above few meters start moving. A wind turbine with a shorter tower shall, therefore, generate more power during the day whereas the one on a tall tower tends to produce a greater proportion of power during the night.

Also, landmasses heat faster during the day and the hot air moves upward creating a vacuum which drives air masses from oceans to land. At night, land masses cool faster while the ocean is still hot which reverses the direction of wind flow. Difference between daytime and night time temperatures vary during different seasons and therefore the pressure differential shall be different which accounts for the varying wind speeds at different seasons.

Wind normally flows from high to low-pressure zones. At latitudes between the subtropics and Polar Regions, wind direction is impacted by Earth's rotation. In the northern hemisphere, the wind rotates in a counterclockwise fashion generating cyclones. Circular anticyclonic patterns appear in the southern hemisphere.

Variability of wind makes it difficult to produce power at a sustained rate as compared to a gas-driven turbine which can be started and stopped at will. At any selected site, power output may be zero for up to 20% of the time and intermediate for the remaining 80% time. The prediction of power in the interday range is poor as compared to prediction over a long period of time like a month or a year. Since it cannot be switched on and off like a conventional power plant, the excess of energy produced at any stage needs to be stored in some form to be used later.

9.4.3 STORMS

Wind speeds exceeding 90 km/h are commonly referred to as storms. Storms are extreme events and can last from 12 to 200 h, depending on the season and the part of the world. Storms arising out of the eastern and northeastern winds in the United States frequent with regular periodicity occurring especially during the cold period. Terrestrial storms alter the oceanographic conditions which affect the deep-water currents which impact nutrient mixing and aquatic habitats. Terrestrial storms produce strong currents, changed tide levels, increased sediment transport, lateral and vertical thermal mixing. Storms are created when a center of low pressure develops with a system of high pressure surrounding it which creates a cyclonic combination of wind and clouds. The scale of this configuration results in the severity of its impact. Storms have enough power but their unpredictability in terms of location and severity makes it impossible to harvest energy.

9.5 WIND ANALYSIS

Wind being a vector is characterized by its direction and speed and is subjected to changes in topography and weather. It is not impossible to estimate wind direction and speed at intervals for which it is not recorded at weather stations, however, any such estimation is not free from error. There are several mathematical functions called probability density functions that can be applied to model the wind speed frequency curve. In wind power studies, Weibull and Rayleigh probability density functions are commonly used and widely adopted (Patel, 1999).

9.5.1 WIND POWER

The blades of a windmill capture the kinetic energy of blowing wind and make it rotate. This turns the potent energy in wind into mechanical (rotational) energy. This rotational energy can be directed into driving an internal shaft connected to a gearbox to amplify the rotational energy. The gearbox can be used to achieve some productive work like lifting water, grinding food grains or attached to a generator to produce electricity. In order to estimate the power or productivity of a windmill, wind patterns from speed measurements are collected and analyzed. The design of the windmill or turbine is adjusted to rotate to face the strongest wind and angle or "pitch" its blades to optimize the energy conversion.

When wind flows with a velocity *v*, then the kinetic energy (*P*) available is given by the below equation:

$$P = \frac{1}{2}mv^2 \qquad (9.1)$$

where *m* is known as the mass flux (flow rate) and is denoted by:

$$\frac{dm}{dt} = \rho A v \qquad (9.2)$$

where *p* is the density of air, *A* is cross sectional sweep area, and *v* is velocity. Combining the above, we arrive at the below equation for wind power:

$$P = \frac{1}{2}Av^3 \qquad (9.3)$$

Hence power from wind is dependent on wind density, sweep area, and wind speed. Wind speed is dependent on geographical location and distribution of landmasses which makes some locations more suitable for high speeds than others. Density is temperature dependent hence air at Polar Regions contributes to more power than the tropics. Sweep area depends on the diameter of the turbine blades and longer the blage, higher is the power generation capacity. The performance of a wind turbine is determined by its power co-efficient (*Cp*) and is given by:

$$Cp = \frac{Electrical\ Output}{0.5 * \rho * A * v3} \qquad (9.4)$$

9.5.2 WIND SPEED AND MAPS

Wind speed varies with time of the day and day of the year and stochastic in nature. Therefore the power produced from wind is also stochastic. Analysis of wind speed is therefore critical for understanding the electrical output possible at a given geography.

Wind speed is measured using anemometers, which is a common weather station instrument. Anemometers can be of various types such as sonic anemometers, cup anemometers, and windmill or propeller anemometers. Today we have multifunction anemometers which combine wind speed measurements with pressure, temperature and humidity readings simultaneously (Fig. 9.3). Wind also can be recorded remotely using LiDAR profilers which apart from providing wind speed can also provide information on air turbulence.

Weather stations make use of anemometers for measuring wind speed and direction. Wind data recorded at different weather stations at particular elevation are averaged over a temporal range to produce wind maps (Fig. 9.4). Availability of fast and accessible internet and its increased use coupled with the rich open global datasets which describe the Earth's atmosphere and surface conditions have set the scene for production and usage of wind maps and atlases. Growing concerns about climate change and extreme events have driven many researchers and agencies using these wind atlases, which presents an understanding of wind speed and patterns from regional to global scales. High precision and frequency data are normally used over large periods to analyses the energy harvesting potential by large wind farms.

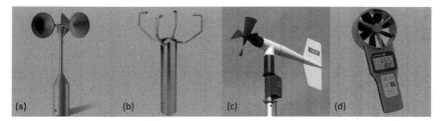

FIGURE 9.3 Types of anemometers. (a) Cup anemometer, (b) sonic anemometer, (c) windmill anemometer, and (d) multifunction anemometer.

Source: www.pce-instruments.com.

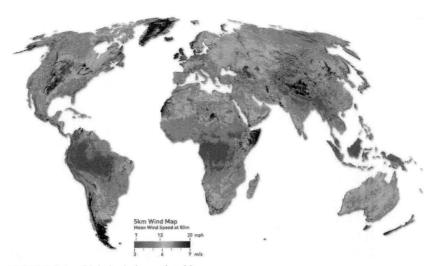

FIGURE 9.4 Global wind speed at 80 m.

Source: 3Tier.com.

9.5.3 STATISTICAL ANALYSIS

Globally, power generation from wind is dependent on topographical and meteorological conditions. At regional scales, wind speed varies between different latitudes, the proximity of land to large water and the presence of mountains and plain areas. Wind speed frequencies are used to evaluate the efficiency of wind turbines using a power curve. The power curve of a wind turbine is a graph which indicates the electrical power output at different wind speeds. Field measurements of wind speeds at the potential turbine installation site are statistically studied to generate power curves to check the suitability of different wind turbines. Where wind speeds do not fluctuate rapidly, anemometer readings for wind speeds can be used to predict power output from various wind turbine designs.

Spatial and statistical variability of wind is interpolated by computer-based models using (atmosphere and land surface physics, some generalized physical) relations between wind speed, topography, and vegetation (Oliver, 2005). This helps in wind resource assessment of a region. Longterm wind measurements with high spatial and temporal frequency are available only for densely populated cities where more number of weather stations exist. This data is interpolated and extrapolated for extended regions that are not significantly far off. Interpolation methods are therefore critical understanding of the wind energy potential of a region.

A mode of representing wind characteristics of regions is obtained by a wind rose. Wind rose is a graphic tool which gives a broad understanding of wind speed and direction variations over time at a location. It summarizes wind variability at a location, with information on its strength, direction, and periodicity.

The wind rose is normally plotted in the polar coordinate system where the frequency of winds over a time period was represented with color-coded or hatched bands for each direction range. Wind rose is a circular map, akin a spider map which shows the frequency of winds blowing from particular directions. The radial length is representative of the frequency of time that the wind blows from a particular direction. This makes each concentric circle representing a different frequency, from zero at the center to a maximum of 100% at the outermost circle. Additional information may be included in form of hatched or color-coded bands on each spoke in the 16 cardinal directions (0°–360°) with respect to north (Nfaoui, 2012).

There are a number of different formats that can be used to display wind roses, one such is presented in Figure 9.5.

Statistical methods use wind factors like mean wind speed, frequency of wind speed, energetic wind speed, power density, and variance of wind speed (Mouradi et al., 2016). Mean wind speed is the time-averaged speed of wind which is averaged over a specified time-interval. Wind speed is normally measured at 10 m over intervals of 1 h.

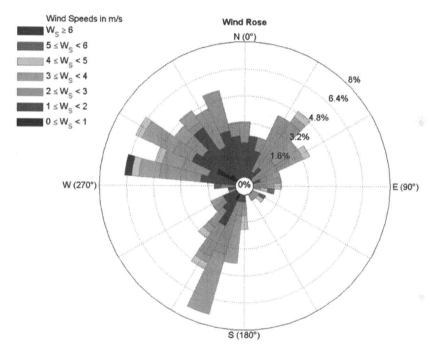

FIGURE 9.5 **(See color insert.)** Wind rose diagram for Changi region Singapore.

Source: prepared with data downloaded from www.windfinder.com.

Frequencies of wind speeds have important applications in environmental analysis and prediction of future wind patterns is greatly helpful in the design and assessment of energy output from wind-powered generators (Mage, 1980). The frequency of wind speed is represented by Weibull distribution which essentially is the wind speed expressed by the probability density function involving the scale factor (c) and the shape factor (a). The Weibull distribution is a standard expression prevalently used for representing the probability density distribution of wind measurements aimed at calculating the mean power achievable from a wind turbine over a given range of mean wind speeds (Fig. 9.6).

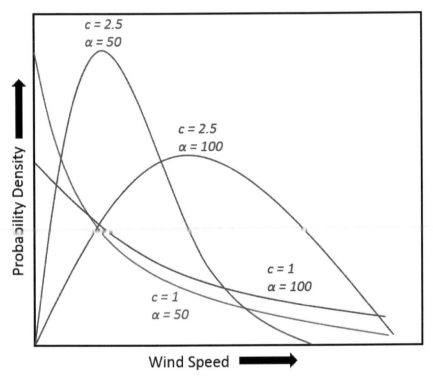

FIGURE 9.6   Weibull distribution curved for different scale and shape factors.

9.6   WIND TURBINES

Wind turbines are devices which convert the potent kinetic energy in wind into electrical energy. Wind turbine vertical and horizontal axis designs involving variable rotor length for different wind zones and landscapes. The smallest turbine designs are used for low load applications like battery charging for small boats or caravans or even traffic signal lighting. Medium sized turbines are used for making contributions to a domestic power supply and the excess power produced can be sold to the marketed via connected grids. Wind farms are arrays of large turbines, installed in suitable landscapes for large-scale power production. Wind farms are becoming an increasingly important source of renewable energy for many countries and feature as a part of a strategy to counter fossil fuels. Wind turbines are classified by the wind speed they are designed for and the plane along which the rotors rotate. Horizontal-axis wind turbines (HAWT) as the name suggests have the blades hinged to a horizontal shaft to which the electrical generator is attached at the top of the tower. Small HAWTs have a simple wind vane; whereas large ones usually have a wind sensor coupled to auto adjust to the direction of wind. Most have a gearbox, which accelerates the rotation necessary to produce the torque requires for the attached generator to produce electricity. Vertical-axis wind turbines (VAWTs) have the main rotor shaft arranged vertically which also serves as the supporting pillar shaft. This configuration has an advantage that the turbine does not require a sensor to detect and orient the axis to face the maximum wind speed direction. VAWTs are suitable for regions where wind direction is highly variable. This design does not require the gearbox and generator to be installed at the same level as the blades. A ground-based gearbox and generator are more accessible for maintenance and repair with less associated risk. VAWTs, however, produce much less energy on a time average, as compared to HAWTs.

9.6.1 LARGE TURBINES

A lot of advancements have been made with respect to wind turbines for large-scale electricity generation. A typical modern turbine is capable of generating usable amounts of power most of the time (>90%). It can start producing power between wind speeds of 4 m/s and 40 m/s. On an annual basis, modern turbines achieve 40% of their rated maximum capacity which is considered superior to conventional electric generation from fossil fuels as unlike the latter wind turbines do not operate round the clock. Commercial large turbines have an individual capacity of about 8 MW and a rotor diameter of between 120 and 180 m (Fig. 9.7).

The installation of such huge wind turbines requires a large amount of space and infrastructure (The Vestas V164-8.0, an offshore wind turbine joins the grid at a cut-in wind speed of 4 m/s, rated wind speed being 13 m/s). Such turbines being site specific, operate at higher ranges of wind speed, and separate transmission and distribution network for buildings/

consumers located far away from point of generation is required. An alternative is the microwind turbine which can operate in urban high-rises tapping low wind speeds.

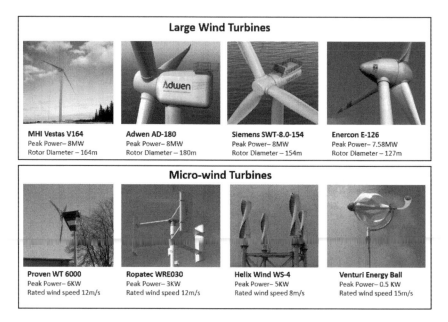

FIGURE 9.7   Some commercial large turbines and microwind turbines.

9.6.2   *MICROWIND TURBINES*

Microwind turbines are small wind turbine meant for microgeneration, as have greater individual power output compared to large commercial wind turbines in wind farms. Deployable at the residential scale, their blades are usually under 3.5 m in diameter and each can produce 1–10 kW of electricity at their optimal wind speed. Micro wind turbines can be very lightweight with some even under 16 kg and are highly sensitive to minor wind speeds (Wagner and Mathur, 2013). Majority of all microwind turbines are horizontal axis type but vertical axis types also found in the small wind market (Fig. 9.7). Urban high-rise roof-tops are ideal for their installation.

Microwind turbine technology with under 1 kW capacity is a relatively new concept intended for building integrated wind generation. Building mounted urban wind turbines (BUWT) have a capacity between 1 and 3 kW and cater to the energy needs of the buildings they are mounted on and are seldom connected to the distribution grid. Building integrated wind turbines serve as an economically viable option and pose as a contender for roof-top photovoltaic installations. They could be installed close to the particular building in order to supply electricity as per its demands, mounted on the building roof-tops, or integrated within the building. Such buildings are often designed in a way to produce the concentrator effect (Dong Li, 2010).

Microwind turbines prove as an excellent source of energy in rural areas due to the laminar profile of wind and low roughness of the surface, their contribution to energy production and CO_2 reduction in the urban terrain is of significance. The major challenge for installation of microwind turbines in an urban environment is the complexity of modeling the wind profile in built-up environment apart from other concerns regarding structural stability, noise levels produced by the turbines, and aesthetic concerns.

9.6.3 CUT-IN AND CUT-OFF SPEEDS

A turbine would be 100% efficient if it converts all the wind's power to mechanical energy. But at low speeds, the wind power does not provide enough torque to rotate the blades of the turbine. The minimum speed of wind required to drive the blades is called cut-in speed and is normally at 3–5 m/s. High wind speeds result in extreme torque which presents a risk of damage to the turbine and the motor. The speed beyond which this risk exists is called cut-out speeds as braking systems of the turbine would bring its operation to a halt. For most turbines, this speed is usually around 25 m/s.

9.7 CHALLENGES AND IMPACTS

Unlike fossil fuels, where nature provided energy over millions of years to convert biomass into energy-dense solids, liquids, and gases—requiring only extraction and transportation technology for us to mobilize them—alternative energy depends heavily on specially engineered equipment and infrastructure for capture or conversion, essentially making it a high-tech manufacturing process (Fridley, 2010). Wind energy like other renewable energy faces the challenge of complementing fossil fuel resources and replacing it completely. Some of the challenges pertinent to wind energy harvesting are discussed below.

9.7.1   INTERMITTENCY AND SCALABILITY

Being dependent on geography, land, and ocean mass distribution, seasons, and day–night temperatures, wind speed is inconsistent and therefore power output is highly variable. Energy service reliability is highly impacted as energy demand as the scale of power production does not match with consumption patterns. At the same time overproduction is wastage and a facility to store energy for future use is required to tap the excess for later use.

9.7.2   ENERGY RETURN ON INVESTMENT

Suitable landscapes for large wind farms are far away from populated cities and therefore the energy produced needs to be transported and integrated with the power grid. A second problem is a difficulty in predicting future power generation from a wind farm due to climate variability. With the fast pace of development of technologies for efficient energy conversion from wind can make immediate past investments unprofitable.

9.7.3   IMPACT ON ENVIRONMENT AND ECOLOGY

Public acceptance of wind energy, like other forms of renewable energy sources, is normally higher than that of fossil fuels. Despite being the cleanest of all renewables, certain adverse impacts on environmental and ecology still exist. The potential impacts result from audible and low-frequency noise, shadow flicker, electric and magnetic fields (EMFs) and solid hazardous waste (Walker and Swift, 2015).

Chronic sleep disturbance is the most common symptom resulting from audible noise emanated from large wind farms although the effect varies from person to person. Some exhibit exhaustion, mood problems, issues with concentration and learning. A major cause of concern about wind turbine sound is its fluctuating nature which many find annoying.

Shadow flicker occurs because of rotating blades under sunny conditions. This casts moving shadows on the landscape and also on buildings resulting in alternating changes in light intensity. Shadow flickers are a nuisance and some have even reported epileptic seizure of patients suffering from photosensitivity. Shadow flicker is characterized by its frequency, which is normally between 0.5 and 1.25 Hz and can be easily predicted based on rotational speed of the wind turbines and the number of blades in the design.

Human exposures to EMFs are more common in society. Power distribution lines which supply households and industry run along every street. Electric substations and high voltage transmission lines which connect between cities and towns are also a source of EMFs. The most immediate source of EMF exposure comes from electrical devices in our own homes, for example, computers, television, food processors, etc. EMFs fall off exponentially with distance and therefore nearfield appliances are more a nuisance than distant ones. EMFs are measured using a gauss meter. Potential sources of EMFs in win power plants are the collection and distribution systems installed above and under the ground. These include the generators, the transformers, and other electrical equipment during commissioning of the plant and transmission lines that connect the project to the electric grid.

Another potential environmental issue arises out of the handling of solid and hazardous wastes generated during the construction and operation of a wind farm. Construction activities for setting up of wind farms degrade the land and produce several solid wastes which pose a hazard to the environment. Packing material and crating materials are nonbiodegradable and they need to be disposed of amicable to avoid their environmental impacts. Activities like land clearing for vehicular access for heavy machinery and personnel alter the land cover exposing the landscape for erosion and sediment movement. Hazardous material like asbestos and lead-based paints requires specialized practices for their handling and removal from the environment after project stage. Compliance with environmental laws must be ensured for shielding the landscape from any potential environmental damage.

Safety to life and property is another aspect that needs to be considered in wind power projects. Depending on the site, the turbine towers may impact aviation safety and interference with telecommunication systems. Large turbines are massive structures and their installation is no less than an urban high rise construction site. Turbine components, concrete and road materials need to be transported to each site in the farm and movement of heavy equipment always are associated with risk to life. Personnel working on the project work at substantial heights and around objects which are rotating and which carry high voltage. Potential accidents like turbine fan blade drop, lightning strikes needs to be addressed in the project safety and accident mitigation plans.

9.8 CONCLUSION

Wind power is constantly on the rise, having exceeded a global capacity of over 450,000 MW. It has emerged as one of the most mature and reliable renewable energy solutions (Kaldellis, 2012). Research efforts in the field of wind energy and its efficiency are in sync with increasing power generation targets of different nations depict that it is set to increase its footprints on global power generation. Options for offshore wind farms have added to its potential. Development of fourth-generation wind turbines and the shift from large-scale rural wind farms to urban high-rise installation for small-scale domestic production has made the technology available to households.

With the deterioration of the environment and depletion of conventional resources, renewable energy has attracted people's attention (Hu et al., 2013). The European Wind Energy Association aims at producing a huge proportion of global wind energy from offshore farms by 2030. If the Global Wind Energy Council is to be believed, the total installed capacity of on both offshore and onshore is expected to exceed 1000 GW by 2030. In order to achieve these ambitious targets, it is necessary that the energy produced from wind is connected to electricity markets. This is an area where several bottlenecks exist, in the areas of design and material, grid integration, and penetration into densely urban consumer bases. Other issues are related to energy efficiency and exploitation of the deeper seas for wind farms. To overcome these challenges, economies need to increase the share of funding support for research and development projects related to wind energy harvesting.

The wind, being a traditional form of energy in the rural sector and having been used since ancient times, has an edge over other forms of energy. However, to make it competitive against other emerging forms of energy and to support its penetration in to newer electricity markets, several policy changes would be required. Subsidies for the adoption of particular technology are important to promote its use and make it popular. Cost reduction in energy production and generating additional income for cities from selling wind power to the local grid is very much desired for its promotion.

Environmental performance of wind farms in impacting local ecosystems is critical to ensuring its sustainability. In other terms, environmental benefits from increased of wind energy need to be more pronounced, for its promotion and social acceptability.

To conclude, wind energy is a mature and reliable power source and has established its importance in electricity production globally. However, its footprint is still very low compared to other alternative sources of energy. The potential is limited by the availability of land and by its cost efficiency and competitiveness. This has constrained onshore wind farms and expansion into offshore farms and their success is the basis for its growth. Advances in micro wind turbine technology are another area which has the advantage of successful deployment in dense urban settings. Despite low capacity, these are low cost environmental friendly and complement well with rooftop solar photovoltaics. Challenges for implementing and promoting wind energy solutions need to be addressed through focused research and development which requires financial initiatives at different scales.

KEYWORDS

- **wind power**
- **wind turbines**
- **windrose**
- **weibull distribution**
- **anemometer**

REFERENCES

Chitnis, A. P. *Source Grid Interface of Wind Energy Systems*. ProQuest Dissertations Publishing: University of Nevada, Reno, 2014.

Dong Li, S. W. A Review of Micro Wind Turbines in the Built Environment. *IEEE*, 2010.

Fridley, D. *Nine Challenges of Alternative Energy*; Post Carbon Institute, 2010. Available at http://www.resilience.org/stories/2010-08-12/nine-challenges-alternative-energy/ (accessed June 15, 2017)

Hu, J.; Wang, J.; Zeng, G. A Hybrid Forecasting Approach Applied to Wind Speed Time Series. *Renew. Energy* **2013**, *60*, 185–194.

Kaldellis, J. K. Wind Energy—Introduction. In *Comprehensive renewable energy;* Sayigh, A., Ed.; Elsevier, pp 1–10.

Mage, D. T. Frequency Distributions of Hourly Wind Speed Measurements. *Atmos. Environ.* **1980**, *14* (3), 367–374. DOI:10.1016/0004-6981(80)90070-0

Mouradia, A.; Ouhsainea, L.; Mimeta, A.; El Ganaouib, M. Analysis and Comparison of Different Methods to Determine Weibull Parameters for Wind Energy Potential Prediction for Tetouan, Northern Morocco. *Int. J. Appl. Eng. Res.* **2016**, *11* (24), 11668–11674.

Nfaoui, H. Wind Energy Potential. In *Comprehensive Renewable Energy*, 2012, Vol. 2, DOI:10.1016/B978-0-08-087872-0.00204-3

Oliver, J. E. *Encyclopedia of World Climatology*; Springer: Dordrecht, New York, 2005.

Patel, M. R. Wind and Solar Power Systems; CRC Press: Florida, USA, 1999.

REN21. Renewables 2017. Global Status Report (Paris: REN21 Secretariat), ISBN 978-3-9818107-6-9, 2017.

Swift-Hook, D. T. History of Wind Power. In *Comprehensive Renewable Energy*; Sayigh, A., Ed.; Elsevier, 2012, pp 41–72.

Wagner, H.; Mathur, J. *Introduction to Wind Energy Systems: Basics, Technology and Operation*, 2nd ed.; Springer: Berlin, New York, 2013.

Walker, R. P.; Swift, A. H. P. *Wind Energy Essentials: Societal, Economic, and Environmental Impacts*; John Wiley & Sons, Inc: Hoboken, New Jersey, 2015.

PART III
ENVIRONMENT

"The race is now on between the techno-scientific and scientific forces that are destroying the living environment and those that can be harnessed to save it…. If the race is won, humanity can emerge in far better condition than when it entered, and with most of the diversity of life still intact."

—Edward O. Wilson

CHAPTER 10

Environmental Impacts of Fossil Fuels

SATYABRATA NAYAK[1,*], MADHUMITA DAS[2], BINAPANI PRADHAN[3], and SANDEEP NARAYAN KUNDU[4]

[1]*Petronas Twin Towers, Kuala Lumpur, Malaysia*

[2]*Vice Chancellor, Fakir Mohan University, Balasore, Odisha, India*

[3]*Department of Environmental Science, Utkal University, Bhubaneswar, Odisha, India*

[4]*Department of Civil & Environmental Engineering, National University of Singapore, Singapore*

Corresponding author. E-mail: satyabrata.nayak@inbox.com

ABSTRACT

Energy demand is peaking due to rapid industrialization and urbanization. Fossil fuels constitute the major source of energy in today's world, which comprises coal, petroleum, and gas from coal seams, shale, and hydrates. Fossil fuel exploration, extraction, transportation, and usage have several environmental impacts. Each stage of the fossil fuel life cycle poses several risks to the environment. During exploration, clearance of land cover aggravates erosion and soil contamination. Drilling for oil and gas can lead to blowouts causing oil spills which are an environmental disaster. Leakage during transportation of oil in subsea pipelines or container vessels also damages the environment. Extraction, transportation, and usage of coal, which is a cheap form of fossil fuel, have several health and environmental hazards. Coal seam gas extraction impacts ground water levels. Shale gas, whose production is now an all-time high, uses fracking which has its own risks for groundwater contamination and microseismic events. Gas hydrates, which are abundant in the seafloor, are a major cause of concern in the wake of global warming as its escape into the atmosphere could aggravate climate change. It is therefore important to have a comprehensive understanding of the potential environmental impacts of fossil fuels and address them through technological interventions making the exploration, extraction, transportation, and usage in environmentally friendly manner for making our Earth sustainable for future generations.

10.1 INTRODUCTION

Use of energy in different forms has been an integral part of human civilization. References to different forms of energy have been reported from different civilization around the world. Dependence on these energy forms is at its peak today because of rapid industrialization and of the many energy resources, fossil fuels have become the most important one. Fossil fuels can be conventional or unconventional, depending on their mode of occurrence and ease of extraction. Conventional fossil fuels are oil and gas which are extracted at ease by means of wells and coal, which has been mined traditionally at various scales. Unconventional fossil fuels include coal gas or coalbed methane (CBM), shale gas, shale oil, and gas hydrates. Many environmental problems have been reported to have arisen from using fossil fuel. Fossil fuels are being condemned to be the reason for irreparable environmental pollution and adverse impact on the ecosystem. Greenhouse effect, acid rain, and air pollution are some of the most studied environmental impacts. The major constituent of fossil fuel is carbon and hydrogen and burning of this fuel produces carbon dioxide (CO_2). Carbon dioxide burnt are causative of the greenhouse effect as they trap the heat inside earth atmosphere. CO_2 concentration and temperature of Earth have seen increasing trend from starting of an industrial age. Apart from CO_2, many other hazardous gases such as SO_2 and NO are formed due to combustion of fossil fuels. The present chapter describes various activities related to fossil fuel exploration and extraction and some of the environmental impacts associated with these activities.

10.2 PETROLEUM EXPLORATION, PRODUCTION, AND THE ENVIRONMENT

Petroleum constitutes conventional hydrocarbons which are explored and extracted in the form of oil and gas. Petroleum is the most sought after fossil fuel because of its high energy potential and ease of storage and handling. Petroleum exploration and production activities have evolved ever since it was accidentally discovered but with increasing demand, its production and usage has increased to unprecedented levels. Present day activities related to hydrocarbon exploration, exploitation, and transportation are performed under stringent environmental and socioeconomic norms. Despite this, we have witnessed several accidents which have been environmentally devastating at various scales. We shall look into some of the activities and their potential environmental hazards which may result from petroleum exploration and exploitation.

Petroleum exploration and production life cycle have four major phases, namely, exploration, development, production, and decommissioning or abandonment (Fig. 10.1). Exploration activities are aimed at locating hydrocarbon in the subsurface. This involves various geophysical and geological data acquisition in the study area and their interpretation which ends with identifying prospects which are oil and gas accumulations that can be drilled and produced. Post discovery of petroleum, several more wells, called appraisal wells, are drilled to determine the extent and commercial potential of the find. Once appraised, the oilfield is developed and production commences, which basically is bringing the resource to the surface from its subsurface abode for usable product making. When production declines and the field is not economical, then the wells are abandoned and the facilities are decommissioned. This is done to secure the environment from any pollution arising out of leakage of hydrocarbons and with the view that the facilities do not pose any hazard to other activities. Petroleum exploration and extraction today is technologically very complex and lapse in procedural compliance at any stage of the cycle has the potential to cause enormous environmental damages. Each stage

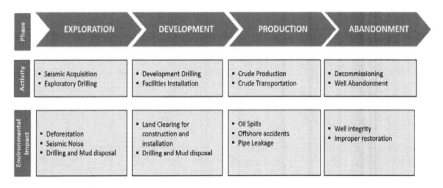

FIGURE 10.1 Conventional oil and gas exploration and related environmental impact.

in the exploration and production life cycle of petroleum presents various hazards which have environmental impacts on land and sea (Fig. 10.2). Details of potential environmental impact of petroleum exploration and extraction which include the likes of seismic data acquisition, drilling, petroleum transportation, and decommissioning of wells are included in the following sections.

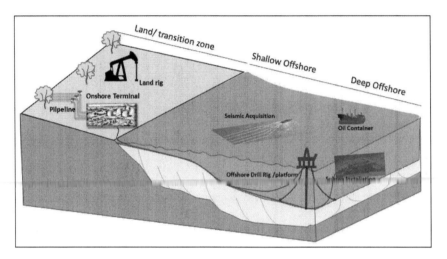

FIGURE 10.2 Schematic diagram showing some of the activities in oil and gas industry which include seismic data acquisition, drilling, and transportation through a pipeline.

10.2.1 *SEISMIC DATA ACQUISITION*

Various kinds of geophysical data such as gravity, magnetic, aerial photography, and seismic data are acquired during petroleum exploration. Seismic data-based exploration is indispensable for the understanding of the structure of subsurface strata and their relationship to a petroleum system. Acquisition of seismic data, on land and on the sea, requires an acoustic source. This acoustic source is produced by explosives or other means which transmit into the subsurface and gets reflected to be detected by sensors. The amplitude and arrival time of the reflected energy is used to image the subsurface.

Previously, dynamite was used as the acoustic source on land which has been replaced by vibrating trucks or "vibroseis" which are relatively low in energy and less hazardous. The receiving sensors are planted in numerous holes on the ground. Seismic sources have the potential to affect society in the vicinity as it produces minor tremors that can damage buildings with shallow foundations. The land-based seismic acquisition also requires the land to be cleared which may need deforestation of forested areas.

Seismic acquisition on the sea makes use of air guns as the seismic source. Death of marine habitat from seismic noise has been reported in many places. These high-energy acoustic waves are damaging to sensory organs of aquatic life forms leading to their exodus from their habitats or even causing deaths. As marine seismic acquisition involves several vessels and towing of streamers which carry the receivers, the likelihood of environmental disaster arising from accidents and spills exist. Experiments have shown that during seismic data acquisition, many fish move to tens of kilometers from the survey area causing a decrease in catch rate (Tsui, 1998). Mass deaths of dolphins which wash ashore in some parts of the world have been attributed to seismic acquisition although the industry has refuted such claims.

10.2.2 DRILLING ACTIVITIES

To test the presence of petroleum accumulation and to produce it to the surface, we need to drill into the rocks. Oil wells often are as deep as 3 km or beyond. Drilling is a critical operation which incorporates high-end, technologically advanced, sophisticated machinery called drilling rigs. Drilling rigs can be classified into many types based on their design and the area they are capable of drilling. Land rigs consist of a mast which supports hoisting of the drill pipes and aids the mechanical rotary drilling. Offshore rigs can be a swamp barge, which operate in shallow depths or a jack-up rig which can operate in depths up to 200 m of water or a drill ship which is designed to drill at water depths of over 3 km. Drilling involves the expulsion of formation fluid to the surface. It also involves mud circulation, cementing where chemicals are used. These chemicals and the activity together pose various environmental hazards.

Rock cutting: Drilling generates huge amounts of rock cuttings. A well of 2000 m depth, with an average diameter of 30 cm, can produce around 150 m³ rock cuttings. These cuttings contain heavy metals which may be radioactive and their production during drilling poses health risks to personnel and people in the vicinity. Disposal of drill cuttings which are also contaminated with drill mud pose an environmental risk. Offshore drilling and seawater contamination with the mud-stained drill cuttings have an adverse impact on an ecosystem.

Drilling fluid: A special mud is circulated while drilling to flush out the rock cuttings to surface. This mud is prepared by mixing chemicals with an oil- or water-based liquid medium. The chemicals may be natural, such as bentonite clay, or synthetically prepared. Continuous circulation of drilling mud through the wellbore cools down the drilling bit, and its density keeps the formation pressure under check.

Synthetic mud should be disposed of with care as they are not environmentally friendly. Pollution due to the mud disposal in marine environments could endanger marine species. Water-based mud or WBM is prepared by mixing salts, particulate minerals, and some organic material with either fresh or brine water. As the additives in WBM are having little toxic characters, it does not pose any serious threat to the environment. Once this fluid is disposed of, mud settles down to form soil. In some types of WBM, heavy metals could contaminate the soil and therefore proper impact assessment needs to be done prior to its disposal. In the sea, WBM may be disposed in seawater and care is taken that it does not end up contaminating the well as its mixing with petroleum can lower quality of crude. It was reported that eighth exploration wells in the North Sea which used WBM for drilling required the disposal of 4000 t of barite and 1500 t of bentonite clay (Lincoln, 2002).

Flaring: In oil wells, gas is coproduced which is often not of commercial value. In the absence of storage and handling of gas, the produced gas cannot be let off to the atmosphere as such and has to be flared or burst into flames. Although flaring is less damaging than letting off the gas into the atmosphere as such, it produces a good amount of CO_2 which is a greenhouse gas. Greenhouse gases are driving climate change and therefore CO_2 emissions need to be contained. A potential remedy is to store gas and use it for secondary recovery or capture the CO_2 from flaring and sequestrate it into the ground for storage which also shall aid enhanced oil recovery.

Wastage: Drilling or production activities are highly sophisticated and technology-driven work where hundreds of people and huge machines work round the clock for few days to few months. These activities produce lots of waste material which are both organic and chemical in nature. Despite a strict rule, not to dispose them into the environment, they are often done so owing to cost and time constraints and poor governance of the regulating authorities.

Water production: During oil and gas production, a huge quantity of water from the subsurface formation is produced, and the amount increases toward the later stages of the well. Formation water could be rich in minerals which contain heavy metals that are radioactive. Reinjection of this water into the formation is the best way to dispose of it.

Blowouts: Blowouts are the most hazardous accidents that can ever happen during drilling. It can be fatal to life and also causes flooding of the surface in the vicinity with crude oil spillage. Blowouts occur when a well is unable to control subsurface pressure. One of the most catastrophic blowout occurred in Mexican coast in 1979–1980, where an estimated 475,000 t of oil spilled into sea (Jernelöv and Lindén, 1981). The most recent blowout in Gulf of Mexico in the United States had a disturbing impact on ecology and marine life which still continues to impact the East US coastal ecosystem and habitats.

Vessel movement: To support drilling activities, the supply of food, etc., for people on board the vessels needs to be transported from onshore to the drilling platform. Such vessel movement is also prone to accidents which cause pollution.

Production platforms: For producing oil and gas, fixed platforms are erected at the offshore site. The possible pollutants from platforms are similar to those from drilling activities of which the subsurface water forms a larger part.

Pipe line: To transport oil or gas from offshore well location to onshore treatment plant, long underwater pipelines are constructed. The pipelines could be damaged due to natural factors such as earthquake, tsunami, etc., or human activities such as ship anchors or any other underwater activities. Depending on the size and nature of the damage, it can leak small or big amount of oil into the sea. Hence, in the first place, the pipelines should be planned in such way that they do not intervene or affect coastal sensitive zones. More than 1300 t of oil spilled in Guanabara Bay in Brazilian coast in the year 2000 due to pipeline leakage (OilWatch, 2002). Similar incident occurred in 1998 in Nigeria, when more than 14,000 t of oil dispersed.

10.2.3 ABANDONMENT AND DECOMMISSIONING

Once the hydrocarbon production in a field is finished, the installation needs to be taken out which is called decommissioning. The decommissioning operation could have many potential impacts on the environment as these involve movement of machinery, pipelines, subsea installations, cutting piles, etc., which if not handled with a plan, could contaminate surface water and groundwater sources. Such offshore activities keep a large zone inaccessible to other normal activities, such as fishing, which also impact socioeconomic conditions.

10.2.4   TRANSPORTATION OF OIL AND GAS

Pipelines are one of the most efficient and cost effective ways of transporting oil and gas from production platforms to the onshore terminal or to the refinery. Leaking of oil pipeline is relatively higher with the spill of oil ranging from few gallons to hundreds of thousands of gallons of oil. These oil spills can affect sensitive ecosystem within which they are running, also affecting human settlement nearby. Tens of oil spills related to the leaking of oil pipelines are reported from the United States every year. From the onshore terminal, oil is transported by railroad. Oil needs to be transported to the refinery and its products from the refinery to the consumers. All these various modes of transporting crude and products are subjected to accidents arising from natural and man-made hazards. Potential hazards are pipe burst or damage, rail wagon derailment, and road vehicular collision accidents.

A severe accident worth mentioning is the 2005 incident in Rzhev where 24 wagons containing crude oil overturned due to derailment causing spillage of more than 300 t of petroleum products. The spillage ended up mixing with Vazuza River, contaminating its waters and triggering an environmental disaster to humans, plants, and animals alike.

In the sea, crude is transported in large vessels over continental distances. Sea transport is cheaper, but nevertheless, not free from accidents. As the sea covers vast extents, any spill can reach the surrounding coasts cutting off water–oxygen mixing which is vital to aquatic flora and fauna. Oil in seawater is toxic to many organisms and when it reaches the cost it spirals a series of environmental and ecological problems. We have witnessed several offshore oil spills in the past, some of which are mentioned in brief in the following sections.

Exxon Valdez spill (1989): This is one of the largest oil spills in the waters of the United States. The tanker Exxon Valdez, which belonged to Exxon Shipping Company, ruptured its hull as it grounded on a reef in Prince William Sound, Alaska. Around 11 million gallons of crude oil spilled into water affecting marine and the costal environment in a huge area. Many creatures such as sea birds, turtles, otters, and harbor seals were found to be affected by the oil spill with various short- and long-term effects. Billions of dollars were spent for the cleanup and restoration of the environment.

Torrey Canyon spill (1967): One of the worst oil spills in the UK coast occurred in 1967. A supertanker named Torrey Canyon carrying oil for British Petroleum struck on Seven Stones reef between Cornish mainland and the Isles of Scilly on March 18, 1967. An estimated 25–36 million gallons of oil spilled into water affecting thousands of square kilometers of area in the UK and French coast.

SS *Atlantic Empress:* This was a Greek oil tanker which collided with another oil tanker, Aegean Captain, in the Caribbean water. The collision which is believed to have happened due to bad weather resulted in the spilling of more than 60 million gallons (287,000 t) of crude oil into the Caribbean Sea.

10.3   COAL AND ENVIRONMENT

Out of the various fossil fuels, coal is being widely used to generate energy for a quite long time now. Coal is found in various sedimentary basins of the world and is used to generate electricity, heat, gas, and diesel. Coal-based thermal power plants are vastly distributed throughout the world. Although coal is regarded as one of the cheapest sources of energy, its use for power production comes at high costs in terms of its adverse impact on environmental pollution and human health hazard.

Coal is extracted from the shallow subsurface through the following mining activities:

i)   open-cast/strip mining and
ii) underground mining.

Open-cast mining is a surface activity where land is cleared and exca-vated to extract the coal which occurs at shallow depths in the subsurface (Fig. 10.3). Where coal is found at greater depths, mining by open-cast methods are not technologically, logistically, and economically viable. Hence, the option of underground mining is exercised, where the main mining activity is to drive a vertical or inclined shaft to access the resource and excavate the coal, hauling it to the surface. Both open-cast and under-ground mining impact the environment in various ways.

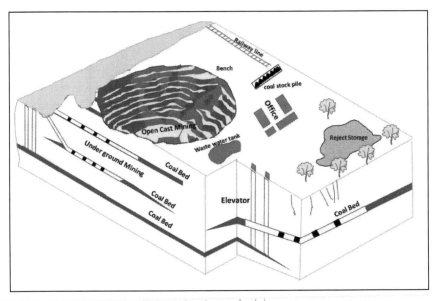

FIGURE 10.3 Schematic diagram showing coal mining and related activities.

In open-cast or strip mining plants, shrubs, vegetation, and top soil are removed to extract coal which lies close to the surface. This process of mining leads to the destruction of forest, landscape, and wildlife habitat and increases the chances of soil erosion and obliteration of agricultural land. Once mining is over, the land needs to be restored to its original form; else it renders the landscape barren. In the United States, coal mining from 1930 onward until 2000 has altered more than 2.4 million ha of forested land. Coal mining in China is reported to have created more than 3.2 million ha of barren land. In Indonesia, coal mining has polluted the environment and reduced the water table, endangering the swamp forests and habitats of endangered species.

Mining activities affect surface water as well as groundwater quality. Waste rock material is also excavated with the coal and is dumped in the open space. Rainwater leaches the chemicals in these rock wastes, producing toxic solutions which contaminated surface water and groundwater. Strip mining, which is a form of open-cast mining, lowers the groundwater levels, which affects the availability of water for the local people. Land clearing for mining reduces the resistance of the land to erosion and renders the land unsuitable for agriculture. Coal particulate contaminates both soil and water, altering their agricultural productivity and human use. Various activities associated with coal exploration, mining, transportation, and power generation are summarized in Figure 10.4.

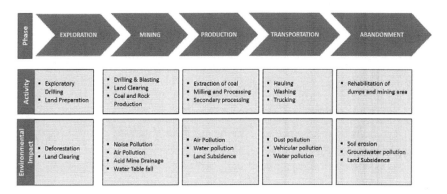

FIGURE 10.4   Various activities in coal exploration and production and environmental impacts.

10.3.1   ENVIRONMENTAL IMPACTS

Underground mining causes many damages to the environment in various ways. Waste materials derived from mining are dumped or piled at the surface creating runoff, which pollutes and alters the local water bodies or streams. Elevated portions of total dissolved solids, sulfates, calcium, and carbonates are also reported from affected water bodies. Underground mining also has impacts on groundwater which includes contamination and physical dislocation of aquifers. Some of the critical environmental concerns associated with underground coal mining are discussed below:

Acid Mine Drainage (AMD)

When coal seams having abundant pyrite are exposed to water and air, sulfuric acid and iron are produced. The acidity in runoff water that dissolves manganese, nickel, and zinc makes the water more polluted and unusable which is toxic for aquatic life. Some metals bioaccumulate in the food chain and impact more animal life.

Mining of coal and its processing produces sulfide, which causes AMD. Bacteria which feed on these sulfides make the water more acidic by decomposing more metal ions, eventually increasing its toxicity. AMD has long-term environmental impacts which do not fade until decades after mining activities cease. Acid drainage percolates into groundwater systems contaminating water produced from wells which are also found to contain heavy metals and other toxins.

Contamination of drinking and industrial water; irreversible impact on growth and reproduction of aquatic flora and fauna; skin diseases; and negative impact on fishery and agriculture industry are some of the many reported environmental impacts of AMD. AMD has been reported from many mining areas of the world with many studies focused on the possible extent of pollution and relating those to the source cause of pollution. A study near around Johannesburg area, South Africa reports that water flow through old mining areas (coal, gold, and other minerals) exposed rocks introduce millions of liters of AMD into the main rivers causing unprecedented pollution. AMD is also reported from various mines of North–East coalfield of India (Rawat and Singh, 1982). The study correlated a considerable amount of AMD in south Kalimantan, Indonesia to abandoned coal mines. AMD is also reported from coal mines of Pennsylvania, Kensington mining of South East Alaska, where the pollution could be related to anthracite coal mining.

Land Subsidence

Underground mining and excavation activities create a void in the subsurface and the land sinks under lithostratigraphic pressure. Land subsidence has even led to collapsing of surface, forming sinks. After completion of excavation in underground mines, the blank space is filled with sand and other rock material which compact under the surface pressure causing subsidence. Subsidence affects surface structures such as road, pipelines buildings, and other establishments. Subsidence over an underground coal mine depends on various factors such as thickness of coal seam, overburden thickness, and mechanical properties. It has been found that subsidence over mines is of the order of 1–2 m and may go up to 3 m where mined coal seams are thick (Holla and Barclay, 2000). This problem can be checked to some extent by preventive methods, such as backfilling of the mine with the waste and combustion waste or leaving some coal in mine to provide structural support to mine roof. Backfilling also carries the risk of contamination of groundwater.

Significant land subsidence due to underground coal mining has been reported from Jharia region of India. Jharia is at the heart of a major coal belt with coal mining in the region could be traced back to 100 years. More than 40 km^2 of area is being found to be subsiding related to underground coal mines (CMRS, 1991). In southern coalfields of New South Wales, land subsidence related to underground coal mining is found to be up to 1.8 m (Jankowski et al., 2008). Similar cases of land subsidence are being reported from Pittsburgh, Scranton, and Wilkes-Barre, which are correlated to underground coal mining in Pennsylvania region.

Coal Fires

Fire in coal (open-cast as well as underground) mines can cause havoc in terms of safety and environmental impact as coal is easily combustible (Krajick, 2005). Forest fires, peat fires, lightning, and other self-igniting processes can cause spontaneous combustion which may lead to coal fires in the mines. Fire in coal seam can actually last for a long time (e.g., years) and can release noxious gases, such as carbon monoxide, CO_2, CH_4, and SO_2, and fly ash into the environment. Various toxic gases resulting from the coal fire can have adverse impact on environment. Continuous fire for a longer time could also result in a rise in temperature in the region.

Coal fire has been reported from various parts of the world with maximum cases in China, where it is believed that around 20–200 million t of coal burns each year (Rennie, 2002). Major coal fires, numbering more than 100, are reported from the Xinjiang province of Inner Mongolia and Ningxia provinces of China. Underground coal fire in Burning Mountain or the Mount Wingen in Australia is reported to be continuing for hundreds of years. Smoke emanating from underground could be seen throughout the region.

Coal Transportation

Transportation of coal from the mines to other places are done using trucks, rail, barges, or through conveyor belts. All of these activities have a direct or indirect impact on the environment. Coal dust blowing from the trucks during transportation contributes to air pollution. Transportation of coal through a pipeline in form of slurry is one of the innovative options which has relatively less impact on air pollution. But, construction of pipeline requires cleaning of forest which has greater impact on local ecology. Again, preparation of slurry itself requires a huge amount of water, which indirectly puts stress on human establishments. The practice of storing coal in open spaces exposed to the atmosphere and precipitation causes air and water pollution.

Coal Processing

Coal washing processes have resulted in the generation of more and more finer particles as the removal of pyrite and associated minerals require finer grinding. Washing of coal has gained much more importance recently, as a result of the increasing environmental concerns and to provide cleaner coal for subsequent use. Efficient flotation, which is based on the differences in the surface chemical characteristics of coal and mineral particles, has become very important. Coal is a complex heterogeneous material composed of a variety of organic constituents in different forms. Its surface properties, such as wettability and floatability, can vary to a large extent due to its complex nature, as well as alterations in situ and through subsequent exposure to various environments. In coal washing plants, as much as 7000 L of water is consumed to produce 1 t of coal and the produced waste materials are pumped to tailing ponds.

Coal-fired Power Plants and Environment

Coal is used to produce steam in power plants. Combustion of coal in thermal power plants produces 100 million t of particulate coal ash. The process also produces gases which contain a toxic mixture of more than 60 hazardous gases which pollute the atmosphere. Major pollutants from coal-fired plants are: (1) burnt coal residue or ash, (2) hazardous and toxic gases, (3) suspended particulates or fly ash, and (4) released gases that are trapped in the scrubber.

The huge amount of CO_2 produced contributes to the global carbon emissions which drive global warming. In addition to CO_2, large amounts of SO_2 and NO_x are produced, which are primarily responsible for acid rain formation. NO_x is the major contributor to forming ozone and nitrates. While ozone reduces the crop yields and impairs lung functions, deposition of nitrates in water bodies is responsible for the growth of algal blooms which affects the major water diversity. Particulates or dust material releasing from the coal-fired power plant cause many health problems such as asthma, bronchitis, and premature death due to heart and lung disease. Trace elements such as mercury

(Hg), arsenic (As), cadmium (Cd), dioxins ($C_4H_4O_2$), and lead (Pb) are released during combustion of coal. These persistent elements accumulate in the environment and can make their way into humans through various agents. Some emitted mercury is converted into methyl mercury by bacteria which can be more toxic than mercury itself. Rainwater leaches the toxins from the open dumps of ash, fly ash, and bottom ash, which is carried by runoff to the flowing streams, some of which percolates and reached the groundwater.

Fly Ash

Fly ash, also known as "pulverized fuel ash," is a coal combustion product composed of fine particles that are driven out of the boiler with the flux gases. Ash pollution is a major issue associated with coal-based thermal power plants where the emanating particles cause haze and also reduce visibility. Similar issues have been reported from various parts of China where coal thermal plants are abundant. Besides these main environmental concerns, coal mining has got long-term impact on our society and environment in form of potential threats to wildlife and impacting human establishment, land use, and climatic changes. Few critical pollutants and their impact on health are presented in Table 10.1.

TABLE 10.1 Some of the Pollutants Related to Coal Mining and Some of Their Health Impacts.

Polluting substance	Health impact
SO_2 (sulfur dioxide)	Responsible for asthma in children and newborn, with long-term effect on lung functionality
NO_2 (nitrogen oxides)	Affects lung function, causes various respiratory illness
HCl (hydrogen chloride)	Infection of respiratory tract, causes cough and other respiratory related issues
Particulate material	Causes respiratory and cardiovascular diseases. Affects heart function (heart rate and heart attack)
HF (hydrogen fluoride)	Can cause respiratory damages, bronchitis, gastritis, etc.
Cd (cadmium)	Affects lung and kidney function
Cr (chromium)	Regular exposure can lead to renal and gastrointestinal infections
As (arsenic)	Can affect lung, skin, and liver leading to cancer
Hg (mercury)	Affects nervous system and respiratory system
Dioxin	Carcinogenic effect on nervous system, could also affect lung system and reproductive system

10.4   COAL GAS PRODUCTION AND ENVIRONMENT

Coal gas popularly known as CBM is an unconventional source of fossil fuel found to be associated with coal seams. Coalification process produces methane gas which is again stored within the internal surfaces of coal beds. In recent times, CBM exploration and productions have been aggressively done around the world.

CBM is produced from the coal bed by drilling numerous wells into the subsurface. Methane is held up in the coal seam due to the high pressure exerted by associated water which is also present in and around the coal seam. To produce the gas, pressure due to the water needs to be lowered through draw down of the water itself. Hence during the production of CBM, a lot of water is coproduced. Some of the above discussed activities and associated environmental concerns are produced in Figure 10.5 for reference.

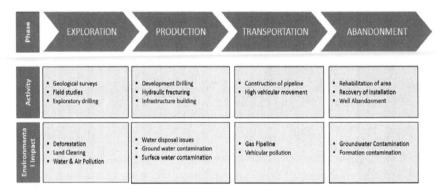

FIGURE 10.5   Vital activities in CBM exploration and exploitation and associated environmental impacts.

CBM is relatively more environmentally friendly than coal as the production of CBM does not involve extensive mining or excavation. Usage of CBM for thermal power plants does not produce wastage, fly ash, and other hazardous gases such as sulfur. Still, there are many environmental concerns associated with CBM exploration and production, a few of which are described below.

10.4.1   WATER PRODUCTION

Dewatering of formations aimed at reducing pressure for the gas trapped in the coal to be released produces huge amounts of formation water for long periods of time. This water is from ancient formations and is likely to be impacted by radioactivity from heavy minerals of the formation it was contained in.

Coproduced water disposal: This water is produced throughout the lifetime of the gas wells and needs to be managed in an environmentally friendly manner. Improper disposal of such huge volume can aid to the erosion of soil or sediment. It can alter local microclimate and impact regional aquatic bodies. The produced water contains particulates and toxins such as coaly matter, ammonia, and hydrogen sulfide (H_2S), which are pollutants. Water disposal could be done through methods of evaporation or percolation where the water is stored in a large peat for a very long time. Although this is a very inexpensive way of disposal, storing the water for long periods could actually facilitate its percolation and eventual contamination with local groundwater impacting its quality. If the produced water is saline, then its disposal in agricultural land will make it infertile and also cause saline water intrusion into local water bodies. Water from CBM wells also contain coal particulates, which if consumed can impact humans by causing health complications such as renal and cancer-related diseases.

Depletion groundwater table: As already stated, a large amount of water is produced during production of CBM. This drawdown of water has a significant impact on the local water table. Withdrawal of water could lead to deepening of the groundwater table which could create problem for human use as well as use by other livestock. Massive withdrawal of groundwater can cause subsidence and trigger microseismic events. Coal seam fire has been reported arising due to CBM production. The drawdown of shallow groundwater is a significant environmental concern associated with CBM production. Withdrawal of groundwater from aquifers during CBM production affects the day-to-day life of human beings and livestock. Additional effects of groundwater withdrawal are land subsidence in the localized region. In a few cases, coal seam fire has been attributed to the production of CBM, which could be due to self-heating characteristics of some of the coal types. Heating of the surface due to the fire and expulsion of steam and other chemical gas can destroy vegetation and affect soils on the surface.

10.4.2 GAS WELLS AND POLLUTION

Methane contamination: The poor integrity of CBM wells during production can result in leakage of methane and its seepage into the rock formation above and reach the surface. Methane is combustible and is a potential fire hazard if concentrated in significant levels. Its escape to the atmosphere is also damaging as it is a greenhouse gas and influences global warming. Leaking methane can contaminate groundwater body as well as soil, affecting vegetation and livestock.

Surface disturbances: Commercial CBM production requires drilling of several wells over short distances. This requires the creation of access to these locations which involves land clearing and construction of access roads, pits, pipelines, power stations, and other necessary installations. Such activities alter the landscape in the region and also disturb human and wildlife habitat. Reports suggest that CBM projects in the Powder River Basin in the United States require clearing and construction activities on 0.3 acres of land per well. Considering hundreds of wells which are drilled, several hundred acres of land is disturbed in the process.

Noise pollution: A number of heavy machinery and compressors are used round the clock for drilling and pumping gas and for other activities. Heavy machinery is a major source of noise pollution. The noise levels of a compressor can reach up to 3–4 miles, disturbing habitats within range. Movement of various vehicles associated with the project also contributes to the noise pollution in the area. Installation of noise abatement structures or mufflers is needed in many cases to reduce the impact of noise pollution to a certain extent.

Air pollution: Carbon dioxide emanated from vehicles, compressors, and processing plants, contributes to the air pollution in the area. CBM processing plants emit various hazardous gases such as carbon dioxide, methane, SO_2, and particulate material which have a continuous impact on air quality.

Other issues: Drilled wells into the subsurface are cased with steel casing and are cemented to stabilize it and to ensure sealing of the well from leakage of gas. Despite this, it has been found that most wells leak at some point in the production timeline of the well and release methane in the process. CBM projects could affect the aesthetics of the region, alter the land use, and impact the quality of human and animal life in the region.

10.5   SHALE GAS PRODUCTION AND ENVIRONMENT

Natural gas trapped in extensive shale formations is yet another unconventional fossil fuel resource. Recently, shale gas production has reached new highs, with most of the world's production coming from the United States. The energy potential of shale gas and trends in shale gas production are revolutionary and are transforming its exploration in many parts of the globe. Hydrocarbon production from shale is different from conventional oil and gas production. Fine-grained shale acts as a reservoir for light hydrocarbons in its pores. The fracturing of shale releases oil and gas which then could be retrieved. A complete cycle of shale gas project could be divided into exploration, appraisal, and production and decommissioning. Although in recent time shale gas or shale oil is found to be very much economical, these activities are also associated with many environmental concerns. Some of the major concerns (Fig. 10.6) associated with shale gas/oil exploration, development, as well as extraction are described in the following sections. Another schematic diagram depicting generalized activities in shale gas/oil industry is shown in Figure 10.7.

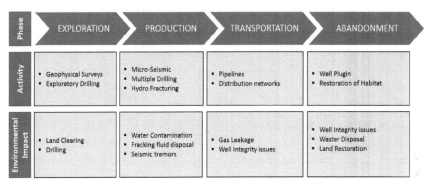

FIGURE 10.6 Summary of activities in shale gas exploration and their environmental impacts.

10.5.1 LAND USE CHANGES

A considerable amount of land is required for construction, infrastructure installation, access roads, pipelines, and other facilities in any shale gas project. In the past, a number of vertical wells are drilled to produce gas, which requires a larger land footprint. But in recent time with the advancement of technology, horizontal well drilling has reduced the number of wells needed, thereby lowering the area impacted by drilling-associated environmental risks.

The horizontal wells can reach greater distances, needing smaller area on the land or surface, thereby reducing the impact at the surface. Few studies report that land use with horizontal wells could be 10 times less than that of vertical wells. The impacts are in terms of deforestation, destruction, and fragmentation of wildlife habitat, as well as dreadful impact on agricultural land and tourism. Although with new technological advancement the impact on land use is being reduced significantly, still utmost care should be taken in planning the drilling campaign in such a way that smaller parts of a sensitive ecosystem are not affected.

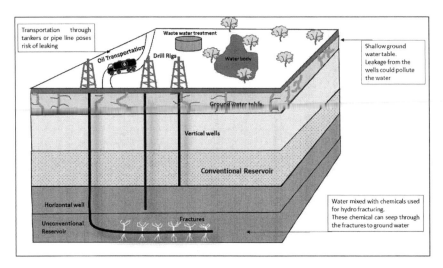

FIGURE 10.7   Schematic diagram showing activities related to unconventional shale gas/oil industry. (adapted from Vengosh et al., 2014)

10.5.2   *HYDRAULIC FRACTURING AND ENVIRONMENTAL ISSUES*

Hydraulic fracturing is one of the most important and critical activities in the shale gas campaign, which is required to generate fractures within the low permeable shale formation. In this process, a number of vertical and horizontal wells which could run across thousands of meters are drilled and then water mixed with sand and few other chemicals is injected into the holes. A considerable amount of water is used in hydraulic fracturing which if leaked to the surrounding formation will adversely affect the groundwater which is used for drinking. Hydraulic fracturing fluids comprise mostly sand and water with small amount of chemicals. The chemicals used in the fluid contain acids, polyacrylamide, ethylene glycol, sodium chloride, isopropanol, sodium carbonate, sodium chloride, etc., some of which are toxic in nature. Although in vertical and horizontal drilling the groundwater is confined by protective casing and cement, still, there is possibility of water leakage through the unsealed casing and poorly cemented zones. Proper management and planning of completed wells and fracturing activities could minimize groundwater pollution to very minimal level. Chemicals and wastewater from hydraulic fracturing if accidentally released, would harm the environment by contaminating or altering the quality of water (AMRA). Hydraulic fracturing also carries the risk of upward movement of gas and saline water from a leaky well casing, as well as natural fractures in rocks may affect the aquifer water quality over long time scales of leaching.

Seismic activity: Studies have shown that drilling of subsurface wells and injecting a large quantity of water in shale gas exploration and development is responsible for generating seismic activities in many regions (IGU, 2012). Seismic activities or minor earthquakes have been reported from the United States and United Kingdom, which are understood to be resulted from injecting water into wells. Whether there should be a seismic activity because of water injection also depends upon the local geology where these activities are being done. Drilling or water injection through shallow fault could actually reactivate the faults and create earthquakes.

Air pollution: During production of shale gas other pollutants such as aromatic hydrocarbons are emitted. Silica and ozone are also released to the air. As stated earlier, extensive drilling activity and other related activities use heavy machinery, drilling rigs, and compressors, which contribute heavily to environmental pollution by emanating CO_2 (Marsters et al., 2015). However, from some studies, it is reported that the life cycle of greenhouse gas emission from shale gas is longer than that originating from coal. Leakage from wellhead and transportation of gas also add methane to greenhouse emission.

Exponential growth has been reported in unconventional oil and gas exploration sector around the world. Although new environmentally friendly technology is widely being used in unconventional exploration and production, the environmental impacts of these activities still need to be studied with utmost importance to avoid any potential disaster.

10.6 GAS HYDRATE AND ENVIRONMENTAL CONCERNS

Gas hydrates are ice-like features made up of water and methane (gas). Gas hydrates are stable within certain depth range below the sea bottom, with the range varying from basin to basin. Various studies have reported the presence of potential gas hydrates in various offshore regions in the world. Although gas hydrate could be an alternative energy resource, it is still in developing phase with not many significant commercial productions yet. Several studies and research are going on for gas hydrate production. Some of the critical activities associated with gas hydrate exploration and potential environmental impacts are shown in Figure 10.8. A schematic diagram showing gas hydrate and the exploration activities are shown in Figure 10.9. Various technological advancements in the recent times are aimed at producing hydrates from the subsurface. The process of exploration and production of gas hydrate has various potential environmental concerns.

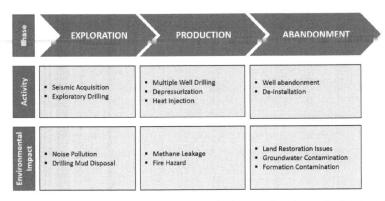

FIGURE 10.8   Critical activities related to gas hydrate exploration and related environmental impacts.

10.6.1 OCCURRENCE

Investigation through deep sea drilling projects and national gas hydrate projects found that a number of wells have been drilled in various sedimentary basins around the world to investigate potential presence of gas hydrate. In this process, presence of hydrate or associated bottom simulating reflectors (BSR) is identified from seismic data and then wells are planned to test the features. Drilling results have confirmed the presence of gas hydrate in many sedimentary basins around the world.

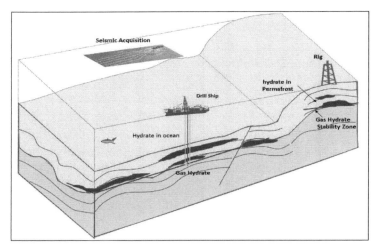

FIGURE 10.9   Schematic diagram showing gas hydrate exploration activities which include data acquisition and drilling of well. Gas hydrates could be found in offshore or in permafrost region. Gas hydrate stability zone refers to a zone or depth in the marine environment where hydrates exist naturally. Red color lines represent BSR or bottom simulating reflectors. Brown color filled polygons represents possible hydrates.

10.6.2   ENVIRONMENTAL RISKS

Impact related to operational activities: Gas hydrate extraction included processes such as heating, chemical injection, and reservoir depressurization.

Well control: Although gas hydrates are buried in relatively shallower depth than conventional hydrocarbons, these act as pressure pockets in shallow sediments where they lie trapped in the solid state. Exploratory drilling for gas hydrates, therefore, is associated with well control risk which requires utmost care to avoid drilling accidents such as blowouts or rig collapse.

Water pollution: During the production of gas hydrates huge amount of water is coproduced. Production of gas hydrates also requires injection of chemicals to destabilize the solid hydrates, enabling the release of methane. The coproduced water, therefore, is contaminated with gas, oil, and the injected chemicals which require its safe and environmentally friendly disposal to prevent further pollution.

Air pollutions: Drilling of wells for hydrate exploration and production encompasses usage of drilling machinery and compressor station which contribute to air pollution in the region. Any accidental release of gas from pipelines can also be a bigger pollutant.

Methane release and global warming: Gas hydrates buried under sediments are only stable in a tight temperature and pressure condition known as the gas hydrate stability zone (GHSZ). In an area, this zone is presented by various factors such as seabed temperature, water depth, and geothermal gradient. Slight changes in the pressure and temperatures can disturb the stability of gas hydrates leading to their disassociation and release. Brewer (2000) suggested that even a minor sea level change can cause hydrate disassociation and releases methane. In the geologic time, it has been found that methane concentration during glaciation is relatively lower than interglacial stages (Kennett, 2000, 2003). Emission from hydrate is one of the major sources for methane concentration in the air. The concentration of methane in the air is far less than that of CO_2 (about 1/200); still, methane is an important greenhouse gas which is characterized by global warming potential index 3.7 times more than that of CO_2 in mole number and 20 times more by weight (Zhen-Guo et al., 2012).

Submarine mass failure and tsunamis: Dissolution of gas hydrate and release of gas entrapped there can actually lower the mechanical strength of the surrounding sediment or seabed, resulting in submarine mass failure or landslide. Many of the Holocene submarine mass transport complexes or landslides are interpreted to be due to decomposition of gas hydrates. A number of submarine slides or mass transport events in Norwegian continental slope are interpreted to be caused due to hydrate disassociation (Bugge, 1987). One of the biggest mass movements identified off Norwegian continental slope is storage slide which extends more than 800 km with thickness around 450 m.

Ecological risk: Exploration of gas hydrate on seabed sediment may change the marine ecosystem of a particular area drastically and carries the risk of an underwater gas blowout. Expulsion of methane into the seabed can also affect various species in the region.

10.7   CONCLUSION

Fossil fuel has been a major contributor to energy sector for centuries now. There are debates all around the world regarding the negative environmental impact of fossil fuel use. Exploration, exploitation, and transportation of various fossil fuels have a different adverse environmental impact.

Increase in CO_2 and greenhouse gas has been correlated with the extensive use of fossil fuels. Coal is still abundantly used for thermal power plants and other industries because is it abundant and cheap at the same time, which is why shutting down of coal-fired power plants can be devastating to a country's economy. Oil in form of petrol and diesel is widely used as automobile fuel in industry. Transition to electric vehicles is too slow and is constrained by availability of power charging stations and lack of power storage solutions. Hence, ruling out petrol or diesel in the automobile industry is not feasible in the current scenario. The exponential growth of shale gas in recent times has been correlated to environmental pollution in various countries. Horizontal drilling technology in shale gas exploration has significantly reduced the number of wells needed to produce gas and this has lessened the acreage of land needed for planting numerous wells. There is no denying that fossil fuel still forms the backbone of the energy industry. With the absence of any real alternative to the fossil fuel, we need to focus on newer technology which could be used to explore, extract, and use fossil fuel with a lesser burden on the environment. Green and clean energy is a must for better and healthier future. In the present scenario apart from looking for alternative energy, serious focus on the technologies that could make fossil fuel more environmentally friendly is the need of the hour.

KEYWORDS

- coal
- oil and gas
- shale gas
- gas hydrate
- green house

REFERENCES

AMRA (Analysis and Monitoring of Environmental Risk). *Environmental Impacts of Shale Gas Exploitation.* http://www.amracenter.com/doc/ (accessed Oct 15, 2017).

Brewer, P. G. Gas Hydrates and Global Climate Change. *Ann. N. Y. Acad. Sci.* **2000,** *912,* 195–199.

Bugge, T.; Befring, S.; Belderson, R. H.; at al. A Giant Three-stage Submarine Slide Off Norway. *Geo-Mar. Lett.* **1987,** *7* (1987), 191–198 (Bulletin of Oilwatch, 2002).

CMRS. *Surface Subsidence in Mining Areas*; Project Report 1964–11990, Central Mining Research Station (Council of Scientific and Industrial Research): Barwa Road, Dhanbad, India, 1991.

Holla, L.; Barclay, E. *Mine Subsidence in the Southern Coalfield, NSW, Australia*; New South Wales Department of Mineral Resources: Sydney, 2000.

International Gas Union (IGU). *The Facts About the Environmental Concerns*; 2009–2012 Triennium Work Report, June 2012.

Jankowski, J.; Madden, A.; McLean, W. In *Surface Water–Groundwater Connectivity in a Long Wall Mining Impacted Catchment in the Southern Coalfield, NSW, Australia*, Proceedings of Water Down Under 2008, 2008; Causal Productions: Modbury, SA, Australia, 2008; pp 2128–2139.

Jernelöv, A.; Lindén, O. Ixtoc I: A Case Study of the World's Largest Oil Spill. *The Caribbean,* **1981,** *10* (6), 299–306.

Krajick, K. Fire in the Hole. *Smithsonian Magazine*, 2005; Smithsonian Institution: 54ff (accessed Oct 24, 2006).

Lincoln, D. *Sense and Nonsense: The Environmental Impacts of Exploration on Marine Organisms, Offshore Cape Breton*; Submission to the Public Review Commission: Cape Breton Island, Nova Scotia, 2002.

Marsters, P.; Alvarez, F. C.; Barido, D. P. L.; Siegner L.; Kammen, D. M. *Analysis of the Environmental Impacts of the Extraction of Shale Gas and Oil in the United States with Applications to Mexico*; Energy and Resources Group, University of California: Berkeley, 2015.

Rennie, D. How China's Scramble for 'Black Gold' Is Causing a Green Disaster. *The Daily Telegraph*, London, 2002 (accessed Apr 30, 2010).

Rawat, N.; Singh, G. In *Occurrence of AMD in NE Coal Mines of India*, Proceedings, 1982 Symposium on Surface Mining Hydrology, Sedimentology and Reclamation, University of Kentucky, USA, Dec 5–10, 1982; pp 415–423.

Tsui, P. T. P. *The Environmental Effects of Marine Seismic Exploration on the Fish and Fisheries of the Scotian Shelf*; Mobil Resources Corporation, 1998.

Vengosh A, Jackson RB, Warner N, Darrah TH, Kondash A. Critical review of the risks to water resources from unconventional shale gas development and hydraulic fracturing in the United States. Environ. Sci, Technol.2014. (48),8334–8348.

Zhang, Z. G.; Wang, Y.; Gao, L. F., Zhang, Y.; Chang., Liu, C. S. Marine Gas Hydrates: Future Energy or Environmental Killer? *Energy Procedia* **2012,** *16* (2012), 933–938 (2012 International Conference on Future Energy, Environment, and Materials).

CHAPTER 11

Geospatial Technology Applications in Environmental Disaster Management

MAHER IBRAHIM SAMEEN, RATIRANJAN JENA, and
BISWAJEET PRADHAN*

*School of Information Systems and Modelling, Faculty of Engineering
and IT, Centre for Advanced Modelling and Geospatial Information
Systems (CAMGIS), University of Technology Sydney, NSW 2007,
Australia*

*Corresponding author. E-mail: Biswajeet.Pradhan@uts.edu.au

ABSTRACT

Geospatial technologies like GIS and GPS have grown rapidly and widely
utilized in disaster management. This chapter provides an overview of
the roles and applications of various geospatial technologies in landslide,
oil spill, and earthquake disaster management. It explains the use of
each technology in various stages of a disaster management process, for
example, pre-disaster, during disaster, and post-disaster; and provides
the current limitations of such technologies in reducing the impacts of
these disasters. Finally, it presents examples of recent methods applied to
disaster management using different geospatial technologies.

11.1 INTRODUCTION

Natural and man-made disasters such as landslides, earthquakes, and oil
spills result in significant loss of human lives and damage to infrastructures.
Minimizing the impact of disasters is important, which can be assisted
with aerial photographs, satellite images, and other geospatial forms of
data, such as building footprints and road networks. This chapter presents
a review of the role of geospatial information to advance the applications

of disaster management. The objective of this study is to provide a clear background on these applications and the corresponding geospatial data needed for their detection and modeling in geographic information system (GIS) environments. First, a brief introduction to natural hazards and disaster risk cycles is provided. Then, a summary of the geospatial technologies such as GIS, global positioning systems (GPS), satellite and radar images, and laser scanning with focus on their use in natural disaster assessment and modeling, is laid out. Thereafter, natural disasters such as landslides, earthquakes, and oil spills are explained and discussed in detail. Subsequently, the importance of geospatial technology for government and local authorities are discussed as well. Finally, the limitations and challenges of using geospatial data to assess and model natural disasters at their different stages are presented.

11.2   NATURAL HAZARDS AND DISASTERS

Natural disasters are a global concern because they threaten lives and damage properties worth several billions of dollars every year. Often, natural disasters spread to populated areas; in this case, the effects can be extreme. For this reason, studies that attempt to reduce the impact of natural disasters by using remote sensing technologies and geospatial information are ongoing and increasing. A wide range of disasters occur every year. Notable natural and man-made disasters include landslides, flooding, oil spills, earthquakes, volcanic activities, droughts, thunderstorms, and wildfire. The effects of each type of disaster vary according to their magnitudes and geographic locations. Although it is not possible to fully prevent natural disasters, proper planning and public warning can minimize the effects by a great margin. Modern technologies such as GIS and GPS are efficient tools in making proper plans for natural disaster management because they provide needed information at different stages of planning. Additionally, having mutual aid in place is advisable for maximum mitigation and quick rescue efforts. Besides, keeping necessary records such as vulnerable areas and the likelihood of disasters occurring in these areas is very important in managing natural disasters. On the other hand, mitigation strategies are crucial to have local community involvement in disaster planning and management. This will go a long way in educating, informing, and mobilizing the local populace in preparation for disasters at community and household levels.

Although natural disasters cannot be completely avoided, disasters can be properly managed, and the loss of life can be minimized to a great extent through a four-part cycle of mitigation, preparedness, response, and recovery (Fig. 11.1). At the mitigation stage, disaster managers and decision-makers design long-term plans to prevent natural hazards from becoming disasters or minimize their impact as much as possible (Van Westen and Soeters, 2000). This task could be done via structural and nonstructural measures and depends on the type of disaster (Lewis, 2009). For example, the structural measures of flood disaster include creating flood levees or reinforcing buildings, whereas nonstructural measures consist of risk assessment and land-use planning. At the preparedness stage, methods and strategies are designed to manage disaster when it strikes, including developing communication strategies, early warning systems, and stockpiling supplies (Van Westen and Soeters, 2000). The followed stage (response) is responsible for implementing plans after a disaster, which includes mobilizing emergency services, coordinating search and rescue, and mapping the extent of the damage. In the final stage, the damaged areas are restored through rebuilding and rehabilitation.

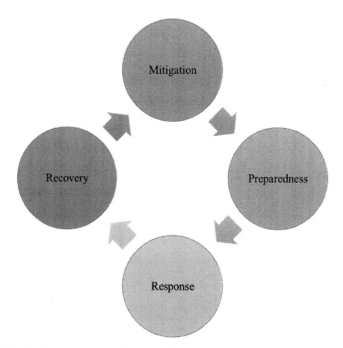

FIGURE 11.1 Four-step cycle of natural disasters.

Geospatial technology has many uses in the four-step cycle of natural disasters from risk modeling, early warning, and damage assessment. Table 11.1 summarizes the use of geospatial technology in the stages of natural disaster management. For landslides, remote sensing techniques such as laser scanning are used to produce high-resolution digital elevation models (DEMs) (Pradhan et al., 2009a; Mann et al., 2012; Jebur et al., 2014b; Van Westen and Getahun, 2003; Li et al., 2016), aerial photography and satellite images for landslide inventory mapping (Pradhan et al., 2016), and a combination of these techniques for hazard and risk assessment (Althuwaynee and Pradhan, 2017). Remote sensors of high temporal resolution and geostationary satellites are used to monitor rainfall and slope stability in the initial stages of a landslide event (De Paratesi and Barrett, 1989). In addition, high-resolution visible images and aerial photography could be used to assess the areas affected by a disaster (Sun et al., 2016; Pradhan et al., 2009). On the other hand, geospatial techniques, such as synthetic aperture radar (SAR) images, are useful for delineating floodplains and laser scanning for flood mapping and hazard modeling (Pradhan et al., 2014). Besides, high-resolution satellite images are often used for land use extraction and quantification of the affected flood-prone areas. Moreover, GIS and Google Maps are effective geospatial tools that are widely used for earthquakes (Ural et al., 2011). These tools are useful in planning routes for search and rescue and evacuation. By contrast, optical satellite images (Janalipour and Mohammadzadeh, 2016), SAR interferometry (InSAR) (Sharma et al., 2017), and DEMs (Turker and Cetinkaya, 2005) are used for earthquake damage assessment, deformation mapping, and hazard modeling. The details of these applications are explained and discussed in the following sections of this chapter.

TABLE 11.1   Use of Geospatial Technology in Natural Disaster Management Stages.

Disaster	Mitigation	Preparedness	Response
Landslide	Digital elevation models Inventory mapping Hazard assessment Risk modeling	Monitoring rainfall and slope stability	Mapping affected areas
Earthquake	Building stock assessment Hazard modeling	Measuring strain accumulation	Planning route for search and rescue Damage assessment Evacuation planning Deformation mapping
Oil spill	Ocean current and wind simulations Risk assessment Inventory mapping	Oil spill detection and oil type classification Trajectory simulation	Mapping affected coastal ecology and marine environment

11.3 GEOSPATIAL TECHNOLOGIES IN NATURAL DISASTERS

Natural disaster education is often needed to view nature and its mechanisms of disaster. Most of the natural disasters occur due to the earth's processes, including morphological expression of the topography within and around the natural disaster areas. The most common remote sensing tool used to understand and monitor natural disasters are aerial photographs and satellite images. Aerial photography has become one of the standard tools aiding the study of various types of natural disasters such as landslides, floods, and volcanic activities. Other techniques include SAR images and light detection and ranging (LiDAR) point clouds, GPS, simulation models and databases, InSAR, stereophotogrammetry, softcopy photogrammetry, digital photogrammetry, and stereoscopic images. This section briefly explains these techniques along with their use in natural disaster management and modeling.

11.3.1 GEOGRAPHIC INFORMATION SYSTEM

A large database is necessary to analyze and monitor natural disasters because the data collected has to be stored, manipulated, and applied. GIS is ideal at this stage of natural disaster assessment because of its capability to handle a large number of past, present, and future data and integrating this data with predictions. Also, GIS is capable of data storage and visualization, is cheaper and easier to use than a manual map production and overlay, and has regional databases, thereby enabling local and regional risk modeling on various types of natural disasters. GIS packages come in various types according to its hardware requirements, spatial functionality, richness of data, and its organization.

GIS provides historical data and information managed in effective large databases from which hazard maps are often generated, indicating which areas are potentially dangerous (Van Westen and Soeters, 2000). GIS is helpful in different stages of natural disaster evolution. During the occurrence of disasters, geospatial technologies and GIS are important for collecting data from air and spaceborne platforms quickly and analyzing them to monitor the evolution of the disaster. GIS is also highly efficient in planning evacuation routes and selecting optimal places for emergency centers through its spatial analysis capabilities and network analysis. In addition, images provided by remote sensors could be analyzed in GIS to assess the affected areas and estimate property damages and losses. Overall, the information collected at different stages of a disaster can be used to update the existing database to make future disaster management more efficient.

11.3.2   GLOBAL POSITIONING SYSTEM

GPS is a method of acquiring three-dimensional (3D) location information on ground or space objects within accuracies reaching a few millimeters (static mode) (Wübbena et al., 2000). Information from GPS can be acquired in real time, allowing better monitoring of natural disasters and designing better early warning systems. GPS is a useful tool for detecting first-stage disaster and further mitigation. Monitors can be placed in any location that can easily be accessed and is relatively easy to operate. GPS units use satellite to determine accurate 3D locations allowing estimation of the mass movements and monitoring them over time. Furthermore, advanced GPS systems, such as differential GPS (DGPS), are often integrated with inexpensive seismic monitors and meteorological sensors to achieve optimum systems for monitoring natural hazards (Riquelme et al., 2016).

GPS, when combined with sensors, such as accelerometers and barometers, can be an important tool for evaluating and possibly predicting earthquakes, landslides, tsunamis, and flash floods. The advantages of advanced GPS include high accuracy, real-time data streaming, and integration with other sensors. However, the use of GPS in natural hazard monitoring has several limitations. These include atmospheric and environmental effects and signal interferences in urban and forest environments. Other limitations of GPS include precision as affected by the number of observable satellites present, obstruction of the observation point, and monitoring of installed GPS receivers placed out in the field. For instance, for Malaysia, with its experience of seasonal monsoon and share of extreme weather-triggered disasters, mitigation measures to reduce risks of landslides are important for the government and the public. They installed state-of-the-art devices in hilly slopes along the North–South Expressway, a major highway in the country, to detect impending landslides and reduce rockfall. The detectors gauge rain on priority slopes and collect data instrumental in determining maintenance requirement on the slope concerned.

11.3.3 SATELLITE IMAGERY/SAR IMAGERY

Several types of optical satellite imagery help to detect and monitor natural disasters. They provide useful information to determine where natural disasters have occurred (detection) or where they are about to occur (prediction). For instance, temporal images can be taken before and after an occurrence of slide movement or flooding events. From these images and simple change analysis in GIS, one can see surface disruption and flood inundation areas. In the extraction of accurate information from satellite images, radiometric and geometric calibrations of the images before and after the disaster are required to make a changed image stand out from those that have not. In addition, optical satellite images, especially those with high spatial resolution, are useful for damage assessment due to natural disasters such as landslides, earthquakes, and floods (Van der Sande et al., 2003).

On the other hand, during rainy seasons, radar images are more beneficial because they are weather independent and can be acquired at daytime and nighttime. Repeat-pass radar images can be applied at different stages of natural disaster management. They can be utilized mainly to detect the movement of active landslides via a technique called SAR interferometry or InSAR (Jebur et al., 2014). This technique can reveal the behavior of landslides, which may not be possible with discrete GPS measurements. Moreover, high-resolution DEM can be extracted from radar images (that is, interferometric synthetic-aperture radar, 5 m), which are very important to produce disaster hazard maps.

11.3.4 STEREOPHOTOGRAMMETRY

As illustrated in Figure 11.2, the method of stereophotogrammetry utilizes sensors mounted on airplanes or drones to acquire two or more images of the same ground scene within a short period so that it can view surface features that have not significantly changed. These images can be processed to obtain topography from the stereo pair of images. The series of stereo pairs offer 3D evolution of mass movements over time and provide useful geomorphic information for other natural disaster applications, such as monitoring volcanic activities (Vassilopoulou et al., 2002).

Although most natural disasters can be monitored with satellite images and SAR techniques, aerial photographs, and stereophotogrammetry are more effective in some cases (Leprince et al., 2008). For example, the study of coseismal deformation using field survey and InSAR is not as effective as that using aerial photograph because field survey and InSAR techniques often fail to provide detailed maps of the near-field surface strain, which may consist of a complex of surface ruptures and cracks within a fault zone of finite width. Therefore, aerial photographs can provide effective approaches to estimate the total slip across a fault zone and its along-strike variability (Leprince et al., 2008).

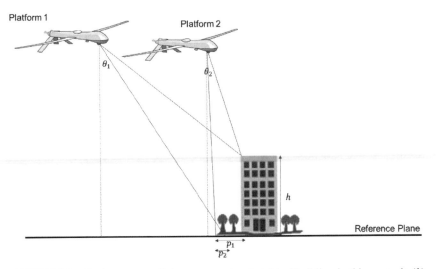

FIGURE 11.2 Basic concept of photogrammetry; height of building in this example (h) can be calculated as $(p_1 - p_2) / (\cot \theta_1 - \cot \theta_2)$.

Several earth observation systems can provide images. SPOT, IKONOS, and WorldView provide stereo images that can be used to produce DEM for the ground surface to monitor natural disasters and surface subsidence. Images acquired by these systems are essential in assessing temporal changes induced by disasters such as earthquakes and landslides. However, the main challenge of these systems is the precise estimation of changes between series images captured by different instruments or those with different spatial resolutions. By contrast, images acquired by sensors mounted on aerial platforms, such as unmanned aerial vehicles (UAVs), can help to overcome most of these drawbacks. UAV systems provide images with very high spatial resolution (<1 m) acquired at different viewing angles and altitudes such that the data can be optimized for various applications.

11.3.5 LASER SCANNING SYSTEMS (LiDAR)

Laser scanning systems are the recent 3D techniques of spatial data acquisition. The word laser defines the sensors that emit a beam (or a pulse series) of highly collimated, directional, coherent, and in-phase electromagnetic radiation. Instruments placed on airborne platforms are called airborne laser scanners (or LiDAR) that offer wide-scale coverage. Figure 11.3 shows the basic concept of airborne LiDAR. The system consists of three components: a GPS, an IMU, and a laser scanner. First, the GPS is used to estimate the coordinates of the sensor while it scans the surface of the earth. Second, the IMU manages the bulk of the positioning workload. In addition to augmenting the GPS in periods of poor satellite coverage, the IMU continually fills gaps between subsequent GPS observations (Williams et al., 2013). Finally, the laser scanner records the point clouds and terrain at certain point spacing. The scanner also records an intensity value, which is a measure of return signal strength that is helpful to distinguish objects of varying reflectivity. In addition to the three main components of a typical airborne LiDAR, digital cameras are often used to collect very high-resolution orthophotos, which aid in better discrimination of the ground features.

FIGURE 11.3   Basic concept of airborne LiDAR operation.

Other forms of laser scanning systems are terrestrial laser scanning (TLS) and mobile laser scanning. These laser scanning systems are efficient for landslide mapping and risk assessments and can acquire dense point clouds used to construct accurate DEM for landslide applications. Data acquired by LiDAR are also used for landslide inventory mapping, prediction, and characterization, simulation of debris flow, and rockfall hazard assessment. The main advantage of LiDAR over aerial photographs and satellite images is its capability to penetrate foliage and acquire data rapidly over large areas. LiDAR has been used for extraction of significant geomorphic, topographic, and hydrologic parameters that support a wide range of landslide, flood, and earthquake applications.

11.4   NATURAL DISASTER APPLICATIONS

11.4.1   LANDSLIDES

Landslides can occur in any part of the world and can result in loss of life, severe damage to property and landscape, and injury to animals. Understanding how a landslide works is difficult because it cannot be stopped. In some cases, landslides result from human activity, but in most cases, it does not. Awareness and understanding of landslide risks are essential in formulating risk reduction strategies, action plans, and contingency plans. The occurrence of landslides led to the creation of many partnerships to implement and enhance the capacity of local authorities on landslide awareness, causes, impact, and operational control in their respective localities. Lack of knowledge and awareness of the dangers it can bring is the main cause of disasters in many areas around the world. Thus, people need to be fully aware of the likelihood of landslide disasters. Increased awareness of the situation may be expected at the government and professional levels. This matter has been given serious attention in the scientific literature, and many emerging technologies have been proposed on the basis of sensing systems and geospatial information.

Landslides usually occur downslope due to the force of gravity. Some occur underwater and are referred to as marine landslides. Over the last few years, a significant number of landslides occurred globally. The force of gravity is believed to be the main cause, although many other factors including human activities have greatly contributed to the occurrence of landslides. One of the most significant forces behind the occurrence of a landslide is the change in slope of a preexisting land making it prone to landslide. Different types of landslides, ranging from shallow landslides to debris flow to rock fall, exist. Geospatial technologies can help in various landslide applications such as landslide inventory mapping, hazard and risk assessment, mass movement detection, debris flow runout distance, and rock fall trajectory simulations. The following sections describe each of these applications including their concepts and the use of geospatial information.

11.4.1.1 *LANDSLIDE INVENTORY MAPPING AND MONITORING*

The term landslide inventory mapping includes all activities aimed at recognizing past landslide events that occurred in a specific region. These techniques allow feeding databases and building inventories for landslides. In fact, landslide inventory mapping involves investigation over large areas. Therefore, a variety of slope instabilities are considered. Usually, all those portrayed in a single map image with a scale of approximately 1:10,000 or above should be used for prediction of such inventories (Harp et al., 2011). Alternatively, an inventory map for each type of landslide can be built. Landslide maps prepared by collecting historical information on landslide events are called inventory maps, which can be further classified as historical, event-driven, seasonal, or multitemporal inventories (Scaioni et al., 2014). Preparation of a landslide inventory relies on the following assumptions. Landslide events leave visible marks on the territory. Thus, visual image interpretation of (stereoscopic) aerial photographs, satellite images, or digital representation of the topographic surface may help the recognition process and the displacement rate of the mass movement.

As with any other natural or man-made disaster that has caused loss of lives and property, the need to study landslides is imperative because they are geological hazards that occur globally. Landslides cause an estimated number of deaths of over 1000 people annually and damages worth over 10 billion USD. The settlement is the first reason for studying landslides because if people decide to settle on potentially landslide-prone areas, they have to understand the risk they are facing. During constructions, the blasting machinery must be considered because this human activity has been found to cause landslides. By studying the science of landslides, how they occur, their triggers, and the general geology associated with them, we can accurately predict and warn people in time, thereby saving lives and property by ensuring quick response via evacuation. The public needs a sense of security from the government regarding natural disasters such as landslides. Thus, the government has implemented programs to ensure a better understanding of landslides and the mechanisms to reduce their impact.

Nowadays, online geospatial techniques, such as Google Earth, provide free 3D models that are employable in the initial stage of landslide investigation or even in extracting information on previous landslides (Fig. 11.4). In addition, GPS receiver and surface visualization using two-dimensional (2D) shaded relief images or 3D perspective view, which are fully integrated within most geospatial systems and remote sensor software packages, have become core tools for the analysis of landscape. Landslide inventory maps are often used as background for further analysis, that is, landslide hazard zonation and risk assessment, given a failure is more likely to occur under the same conditions observed in the past (Devkota et al., 2013). Multitemporal inventories can also be used for landslide assessment. The literature shows that four kinds of remote sensing data are used for landslide inventory mapping, which includes field survey or aerial photographs, optical satellite images, microwave radar images, and laser scanning point clouds (Cheng et al., 2013; Lacroix et al., 2013; Strozzi et al., 2013; Pradhan et al., 2016, Siyahghalati et al., 2016). The conventional methods for building a landslide inventory are field survey and aerial photographs. Stereophotographs are basic tools for landslide detection and are widely applied for topographic 3D mapping. If a couple of stereo images are available over the study area, images can be fused together to exploit the high geometric resolution of the former and the better radiometric information of the latter with a process called pan sharpening. Image visual interpretation is particularly suitable for the back analysis pre- and postevent. Efforts have been made to "automate" remote sensing data interpretation to improve efficiency and help experts during the recognition process. Additionally, the extraction of features or classification of areas where typical signs of landslides are present is a complex task.

Moreover, the optical remote sensor is usually accomplished from airborne and spaceborne platforms and rarely from ground-based platforms. By contrast, microwave sensors are installed on airborne, spaceborne, and ground-based platforms in landslide investigation. The most widely used microwave (or radar) sensor is SAR, an active system capable of recording the electromagnetic echo backscattered from the surface of the earth and arranging them in 2D complex value (amplitude and phase) image maps. The spatial dimensions of such maps are sensor-target distance called "line of sight."

On the other hand, landslide monitoring can help to predict the occurrence of landslides easily through storm forecasting because storms are among the causes of landslides. Scientists in the United States are using landslide hazard program monitors (Staley et al., 2016) to selectively observe several locations that have a high probability of landslide occurrence. This program provides real-time data on the status of the selected sites during all weather conditions. The data received are then reviewed by a committee of experts. When a red-alert situation is detected, the committee advises accordingly. The United States Geological Survey has also used hydrologic monitoring stations in various hillsides of the country. These mechanisms can provide quick and early warnings when destabilization occurs on the hillside, which may result in a possible landslide. Thus, people are evacuated, and properties are saved in time before the disaster hits. These instruments can detect the slightest earth movements, which may indicate a pending landslide. Monitoring using these instruments is reliable in global landslide prediction, thereby saving government resources.

FIGURE 11.4 Google image over Cameron Highlands area in East of Malaysia. This place has experienced many landslides. High spatial resolution and 3D information in the scene allow visual landslide recognition and mapping.

Several studies have presented methods for landslide monitoring based on geospatial technology. For example, Komac et al. (2015) and Bellone et al. (2016) presented GPS systems for monitoring landslides. Komac et al. (2015) integrated GPS with RADAR systems to monitor landslides. According to them, GPS enables detection of abrupt ground motion in 3D and integrating InSAR with GPS/global navigation satellite system (GNSS) provides improved estimates in the height component and estimated velocity of the GPS/GNSS system. On the other hand, Bellone et al. (2016) presented a method for landslide monitoring using low-cost GNSS receivers. Their experiments suggested that instruments and statistical methods in the GPS systems could be useful tools for studying and detecting landslide displacements. A 1-cm precision level (the magnitude of imposed displacements) can be obtained in real time with inexpensive instruments costing a mere few hundred euros. In addition, Xie et al. (2016) presented a comprehensive landslide monitoring study using 3D GIS, and recently, Bovenga et al. (2017) depicted a landslide monitoring approach for risk mitigation based on corner reflector and InSAR techniques using the massive landslide in Carlantino, Italy as an example. Furthermore, Rossi et al. (2016) integrated multicopter drone measurements and ground-based data for landslide monitoring. The result showed that integration of these measurements with data from a wireless network of automated sensing technology could improve the understanding of the movement and reliability of landslide monitoring. Besides, Lian and Hu (2017) used TLS and spatial analysis for landslide monitoring of the Gaoyang coal mine in Shanxi, China. TLS was used to collect point cloud data of the slope landslide, ground steps and cracks, and tilt of a high-voltage tower during underground mining. Their experiments proved to be more effective than the total station in capturing detailed spatial data on high-voltage towers and ground disasters such as landslides.

11.4.1.2   *LANDSLIDE SUSCEPTIBILITY ASSESSMENT*

Landslides have caused massive poverty and loss of property; hence, continuous monitoring and prediction of landslides to avert such losses in the future are imperative. This objective can be achieved by providing accurate information on occurrences of landslides. Methods used for landslide prediction according to the International Scientific Indexing on landslides are available. These methods include landslide susceptibility mapping as a result of remote sensing and geospatial technology. Common methods used for landslide susceptibility mapping include the analytical hierarchy process (AHP) (Pourghasemi et al., 2016; Sangchini et al., 2016), weighted linear combination (WLC) (Michael and Samanta, 2016), fuzzy logic (FL) (Bui et al., 2015), logistic regression (LR) (Raja et al., 2017), support vector machine (SVM) (Hong et al., 2017b), artificial neural networks (ANN) (Gorsevski et al., 2016), frequency ratio (FR) (Hong et al., 2017a), and hybrid models (Tien Bui et al., 2017; Pham et al., 2017).

Landslide susceptibility or spatial prediction of slope failure can be produced by overlaying landslide conditioning factors such as slope, curvature, aspect, land use, geology, soil texture, topographic roughness index, topographic wetness index, stream power index, and vegetation density. These factors are often overlaid after reclassifying them using weights estimated by a regression model (Akgun and Erkan, 2016). Landslide conditioning factors can be derived from several geospatial sources, such as satellite images, topographic databases, and LiDAR point clouds. However, LiDAR was found to be more effective than other data sources. The morphological features of the landslides (e.g., scarps, mobilized material, and foot) are easy to delineate based on hill shades of the produced LiDAR-based DEM. Furthermore, high-resolution DEM and DSM, generated by LiDAR and orthophotos permit production of land use, drainage networks, urban and rural roads, and vegetation structure and density.

The final landslide susceptibility maps are usually reclassified into five classes, namely, very low, low, moderate, high, and very high susceptibility zones. This approach makes the interpretation of the map much easier for decision-making. Several applications for final landslide susceptibility mapping, including land use planning, slope stability management, and landslide risk assessment, are available. Landslide susceptibility is the basic step for landslide hazard and risk analysis. The quality of these maps directly affects the plans set based on the landslide risk assessments. Therefore, producing accurate landslide susceptibility maps for landslide-prone areas is an important step for landslide mitigation. The quality of these maps depends on the quality of the database and the type of model used.

11.4.1.3 LANDSLIDE RISK ASSESSMENT

Computer-based GIS can be used to analyze hazard information and provide national risk assessment data to state and local governments in a quick and easy manner. Specific models can be generated by using the GIS software. In addition, new high-resolution remote sensing capabilities can be examined for use in large-scale risk and vulnerability assessments. Thus, remote sensing and GIS are to be integrated and modeled for the assessment of quantitative landslide-hazard vulnerability. Besides, improvements in monitoring, data collection, and data processing account for most of the advancements made in short-term weather-related forecasting, which helps improve landslide risk assessments. Better modeling capabilities along with a more thorough understanding of landslide conditioning factors and other variables, such as climate change and sea-level rise, are needed to improve long-range forecasting and planning of coastal landslide-hazard impact.

Landslide risk is the expected number of lives lost, persons injured, damage to property, and disruption of economic activity due to a landslide hazard for a given area and reference period (Varnes, 1984). The risk of physical losses can be quantified as the product of vulnerability, cost, or amount of the elements at risk and the probability of the landslide occurrence. Thus, the common method of producing a landslide risk map is by multiplying the landslide hazard (H) with the expected losses represented as the vulnerability map multiplied by the amount of loss ($V \times A$). Schematically, this condition can be represented by the following formula (Van Westen, 2004):

$$\text{Risk} = \sum \left(H \times \sum (V \times A) \right).$$

Landslide risk assessment process (Fig. 11.5) includes several steps such as preparation of landslide conditioning factors, landslide susceptibility mapping, and preparation of landslide triggering factors, landslide hazard, vulnerability assessment, and landslide risk modeling. Many landslide conditioning factors are significant for producing susceptibility maps. The most commonly used factors are the slope, the aspect, the curvature, the land use, the soil texture, and the geology, as well as indices for stream power, topographic wetness and vicinity to rivers, faults, and railroads (Jebur et al., 2014; Domínguez–Cuesta et al., 2017). Landslide hazard is the process of identification and characterization of the potential landslides together with an assessment of their corresponding frequency of occurrence.

In hazard assessment, the probability of landslide occurrence is combined with triggering data such as precipitation and seismic data. In addition, vulnerability is the degree of loss (or damage) to a given element for a given area affected by the hazard. Information on the land use, settlement, and cost of elements are needed at this stage. Landslide vulnerability has been performed by several studies (Winter et al., 2014; Fuchs et al., 2007; Muthukumar, 2013; Romeo et al., 2013; Leone et al., 1996). The term vulnerability is commonly explained as physical, social, economic, and environmental circumstances that can make particular inhabitants highly susceptible to the effects of landslide hazards (Abdul-wahid and Pradhan, 2017). Furthermore, landslide risk can be performed on diverse scales using characteristic strategies for susceptibility and hazard assessment. It can also be subjective or quantitative in structure. Various risks, such as distributed landslide risk, site-specific landslide risk, and global landslide risk, should be addressed in landslide risk assessment (Corominas et al., 2014).

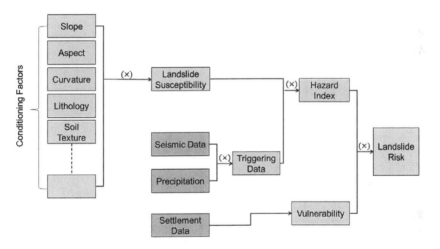

FIGURE 11.5   Typical landslide risk assessment methodological workflow.

Source: Modified from Akgun et al. (2012).

11.4.2   OIL SPILL

Oil spills are another type of natural disaster that threatens coastal marine ecology and environments worldwide. Coastal areas and marine environments contain many species, habitats, and other ecological and biological resources that can be severely affected by oil pollution. In addition, oil often affects the foreshore and increases the damage to the coastal ecology as the spill reaches the beach. Thus, thorough analysis of oil spills via detection and trajectory simulations can improve the preparation for such disasters. This analysis can help assess and predict the endangered coast-line areas and mitigate coastal damage.

Microwave remote sensing, which has the advantages of all-weather and all-day capabilities and capturing images in cloudy weather conditions, is commonly used for monitoring ocean pollution (Brekke and Solberg, 2005). Table 11.2 shows examples of microwave sensors that can be used for oil spill monitoring.

TABLE 11.2   Examples of SAR Sensors Suitable for Oil Spill Detection.

SAR sensor	Mode	Resolution (m)	Swath width (km)	Incidence angle (°)
ERS-2	PRI	30 × 26.3	100	20–26
ENVISAT	Image mode	30 × 30	100	15–45
RADARSAT-1	SCN	50 × 50	300	20–46
RADARSAT-2	Ultra-Fine SP	3 × 3	20	30–49
TerraSAR-X1	ScanSAR SP	1.7–3.49	100	20–45
TerraSAR-X2	ScanSAR	5 × 5	50–500	20–45
ALOS PALSAR-2	Spotlight SP	3 × 1	25	30–44
Sentinel-1	StripMap	5 × 5	80	20–45

PRI, pulse repetition interval; SAR, synthetic aperture radar; SCN, ScanSAR narrow; SP, S polarization.

Other geospatial technologies used for oil pollution monitoring are side-looking airborne radar used to locate oil discharges on sea surfaces (Ferraro et al., 2009), infrared/ultraviolet scanning used to quantify the extent of the oil spill (Brown and Fingas, 2003), microwave radiometer used to estimate the thickness of oil slicks (Pelyushenko, 1995), and laser fluorosensor used for oil type classification (Byfield and Boxall, 1999). Infrared sensors can estimate the thickness of oil slicks because oil absorbs solar radiation and reemits a portion of this energy as thermal energy. In infrared images, thick oil slicks appear as hot objects and intermediate oil slicks appear as cool objects, whereas thin oil slicks are not easily detectable with these sensors. In contrast to infrared sensors, an oil spill can be detected from images if visible under favorable lighting and sea conditions. On the other hand, hyperspectral sensors used for oil spill studies have a potential for detailed identification of materials and better estimation of their abundance (Salem, 2003). Many wavelengths in hyperspectral sensors provide a detailed spectral signature for oil spill slicks, which can be exploited and used to recognize a different type of oil such as crude and light oil.

Figure 11.6 shows the typical methodological workflow applied to study oil spills using geospatial techniques. Several input data, which can be acquired from SAR images, high-resolution satellite images, NOAA's (National Oceanic and Atmospheric Administration) General Bathymetric Chart of the Oceans (GEBCO) bathymetric grid, data collected by geostationary satellites, altimeter radar echo, and Google Maps, are required. These data sources are useful for different analysis stages of oil spill detection and trajectory modeling. In the stage of detection, historical oil spill locations are first collected from oil spill reports for training and validation purposes. For oil spill detection, SAR images have advantages over optical satellite images in highlighting oil spill slicks in oceans. These advantages include the capability to search large areas and observe oceans at night and in cloudy weather conditions. By contrast, limitations of SAR data in the oil spill analysis are the lack of capabilities to estimate the thickness of oil spills and recognize the type of oil. Cloud cover and lack of sunlight limit the use of optical satellite images for oil spill monitoring.

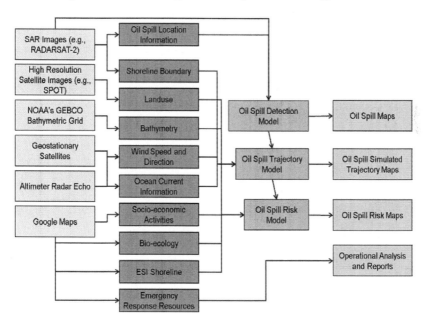

FIGURE 11.6   Overall methodological workflow in detection and trajectory simulation of oil spill using geospatial techniques.

Source: Adapted from Biswajeet and Hamid (2009).

11.4.2.1   OIL SPILL DETECTION

Detection is one of the major parts of oil spill risk assessment and modeling. Visual recognition of an oil spill is not effective because an oil spill can be confused with other substances such as seaweed and fish sperm. In addition, clearly identifying the oil spill through fog and darkness is difficult on the sea surface (Fingas and Brown, 2005). Therefore, geospatial technology is important for accurate and reliable oil spill detection and monitoring. The most common systems used for responding to oil spills are IR/UV and radar. Other techniques used for oil spill detection include laser fluorescence and microwave radiometer (Jha et al., 2008) (Table 11.3). Spaceborne remote sensing has disadvantages of low spatial and temporal resolution, although it provides a synoptic view and a more cost-effective system than airborne platforms. The latter is widely used for oil spill detection and monitoring. Typically, the oil spill detection and monitoring aim to detect the location and spread of an oil spill over a large area to estimate the thickness of an oil spill and determine the type of oil (e.g., crude or light oil). Additional information such as the quantity of spilled oil can also be estimated from this information provided by the geospatial technologies.

TABLE 11.3   Remote Sensing Bands and Related Instruments Used for Oil Spill Detection.

Band	Wavelength	Type of instruments
Radar	1–30 cm	SLAR/SAR
Passive microwave	2–8 mm	Radiometers
TIR	8–14 μm	Video cameras and line scanners
MIR	3–5 μm	Video cameras and line scanners
NIR	1–3 μm	Film and video cameras
Visible	350–750 nm	Film, video cameras, and spectrometers
Ultraviolet	250–350 nm	Film, video cameras, and line scanners

MIR, mid-band infrared; NIR, near infrared; SAR, synthetic aperture radar; SLAR, side looking airborne radar; TIR, thermal infrared.

Source: Adapted from Goodman (1994).

Oil spill detection with visible and thermal sensors has several disadvantages such as the difficulty of distinguishing the oil from the background. In addition, features such as sun glint and wind sheen may create an impression similar to an oil spill; seaweeds and dark shoreline can be mistakenly classified as an oil spill (Jha et al., 2008). By contrast, microwave remote sensing systems have been found useful in oil spill detection for several reasons. First, these systems can work day and night, and because of their side look, can cover large areas in one scene. Second, oil spills appear darker than the background in radar images as the ocean water appears bright because of the reflections of capillary waves (Garcia–Pineda et al., 2013). However, microwave systems also have disadvantages in oil spill detection such as confusion of oil spills with shoreline and calm water. Laser fluorescence is also used for oil spill detection. Laser fluorescence systems record the fluorescence spectrum, which is used to separate gelbstoff and phytoplankton from petroleum oils. In addition, different types of oils have different fluorescence emission signatures, which facilitate oil type classification (Goodman, 1994).

11.4.2.2   SPILL TRAJECTORY SIMULATION

Another important application is trajectory simulation of oil spills. The trajectory of oil spill spread can be modeled in two ways. The first approach is collecting a series of images over the study area at different periods. Once the temporal images are collected, the oil spills are detected using the methods explained in the previous section and mapped in GIS. The changes in the location and spread of the oil spill over time are then used to produce the trajectory of the oil spill in the form of maps in GIS. Another approach is modeling the oil spill trajectory with the aid of additional information on weather and ocean conditions (Marghany, 2004; Price et al., 2006; Biswajeet and Hamid, 2009). This approach requires information on the initial location of the oil spill, which can be obtained from satellite images or aerial photographs. Next, the location of the oil spill is combined with wind speed and direction and current ocean conditions to model the trajectory of the oil for a given period. Both approaches have advantages and disadvantages. The first method requires a series of images over the same study area, which may be infeasible or costly. The second method can be cost effective, but the additional information on weather and ocean conditions can create uncertainties in the model outputs.

11.4.3   EARTHQUAKES

Earthquakes (i.e., shaking or trembling of the ground) are natural phenomena that cause significant surface rupture and severe property damage. Damage reduction strategies, particularly in vulnerable fault zones, are mandatory because of mass deaths and injuries caused by earthquakes over the past decades, urbanization, and rising population. Over 23 million deaths have been caused by earthquakes globally during 1902–2011 alone (http://earthquake.usgs.gov). Figure 11.7 shows the distribution and magnitude of historical earthquakes worldwide from 1965 to 2016. The distribution of the earthquakes is linked to the phenomenon of plate tectonics. Earthquakes cause landslides, damming of rivers, and depressions that form lakes. Thus, earthquake detection, prediction, and damage assessment for the affected places are important research areas.

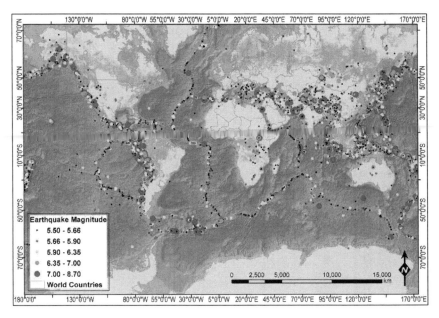

FIGURE 11.7   (See color insert.) Historical earthquakes worldwide from 1965 to 2016.

Source: Map generated from data sourced from National Earthquake Information Center.

Three main seismic belts account for most of the earthquakes to date. These are the Pacific, Alpine–Himalayan, and Mid-ocean Ridge seismic belts where all kinds of earthquake occur, specifically shallow focus, intermediate, and deep focus earthquakes. The Pacific seismic belt is also called the "Ring of Fire" because it is an area where many earthquakes and volcanic eruptions occur in the basin of the Pacific Ocean. The Pacific seismic belt is the largest in the world where the Pacific plate movement is toward the North West side and undergoes subduction beneath the

Eurasian plate. It includes 400–452 volcanoes, which comprise 75% of the world's active and dormant volcanoes. California and the western coastal part of the United States, where earthquakes and tsunamis often occur, lie on the Pacific seismic belt.

As mentioned, a great earthquake can cause extensive damage to human society and environment. The earthquake that occurred in Wenchuan, Sichuan province in China on May 12, 2008, left more than 65,000 people dead. The magnitude of the Wenchuan earthquake was 8.0 on the Richter scale. It was the strongest earthquake in China and had a significant impact on people's lives. China suffered a loss of RMB 845.1 billion (138 billion USD) because of this earthquake. On April 25, 2015, an earthquake with magnitude 7.3 hit Kathmandu, Nepal and left 8857 dead, 8964 injured out of a total of 21,952, and 3.5 million homeless. Total damage amounted to 10 billion USD (approximately 50% of Nepal's GDP). Many earthquakes ranging from 5 to 9 Mw happen in Chile every year, resulting in huge property losses. These cases show that the earthquake prediction is necessary to minimize damages. Observing the strange behavior of animals is a commonsensical way of predicting earthquakes.

To date, many algorithms for earthquake prediction have been developed. The most well known is M8, which predicts earthquakes by analyzing "increased probability with time." Several magnitude scales have been defined to measure the earthquake size, but the most effective and widely used are: (1) local magnitude scale (ML), which is also known as Richter magnitude scale, (2) surface–wave magnitude scale (Ms), (3) body–wave magnitude scale (Mb), and (4) moment magnitude scale (Mw). However, these scales (ML, Ms, and Mb) have limited range and applicability and do not provide a satisfactory measurement of the size of the largest earthquakes.

Geospatial technology is an important tool for the monitoring and prediction of the expected ground shaking as well as earthquake damage assessment. Among the geospatial techniques, GPS is shown to be efficient and better than seismic sensors in determining earthquake parameters such as mechanism and magnitude (Riquelme et al., 2016). GPS instruments can detect the long-period wave directly from displacement measurements in the near field (up to 10°) for large earthquakes. GPS is also effective in calculating the full moment tensor potentially in the nearby field 4 min after each large earthquake. However, in the case of earthquakes with low magnitudes, further improvement of the algorithm and larger GPS data are necessary. In addition, earthquake prediction models are mostly based on accurate GPS measurements. The models are often constructed using parameters such as localization, depth, focal mechanism, and magnitude (Vigny et al., 2005). The model then matches these parameters to detect small surface displacements, which help to infer potential earthquakes. Although GPS networks that are dense and accurate are necessary to measure the small changes of the ground deformations, GPS records are equivalent to very-low-frequency seismograms. They can be used to determine the position of the centroid of deformation at the origin of the bulk of the surface wave emission (Vigny et al., 2005). Furthermore, Colombelli et al. (2013) explored the application of GPS data for earthquake early warning and investigated whether the coseismal ground deformation can be used to provide fast and reliable magnitude estimations and ground-shaking predictions. GPS stations can register the ground displacement directly without any risk of saturation and any need of complicated corrections; geodetic displacement time series represents an important complementary contribution to the high-frequency information provided by seismic data. Therefore, the challenge with GPS data is a practical one and is related to the development of real-time methodologies to retrieve, process, and analyze geodetic displacement time series.

Today, real-time GPS is considered as a valuable component of next-generation near-field tsunami early warning systems that can provide fast and reliable source parameters. Chen et al. (2016) presented a comparison of source inversion techniques for GPS-based local tsunami forecasting using a case study of the Chile earthquake. Local GPS networks with real-time processing can provide ground displacements without saturation almost without delay and are reliable tools for near-field earthquake characterization and tsunami early warning. GPS data are generally useful for real-time monitoring of tsunamis. However, proper data processing algorithms should be used. In addition, O'Toole et al. (2013) and Polcari et al. (2016) presented applications of GPS in earthquake displacement monitoring. The first study integrated GPS with InSAR techniques to monitor earthquakes in the United States. InSAR and GPS data are fully complementary and suitable for integration to estimate an accurate 3D displacement field with millimeter precision and large spatial coverage. The main drawback is the relatively low spatial density of the GPS networks. Typically, the distance between stations that belong to the same network ranges from a few to tens of kilometers according to the wavelength of the signal to be monitored. Therefore, the information provided by GPS often needs to be interpolated; this process might lead to errors that are proportional to the distance between stations so that the entire signal is missing if it falls completely between stations. By contrast, the second study showed that GPS is a useful tool for determining earthquake parameters by measuring the static displacements. They also suggested that a proper algorithm is crucial to the success of GPS application in earthquake monitoring.

A known limitation of high-rate GPS is that its high precision can only be guaranteed in the low-frequency band. For frequencies larger than a few hertz, GPS data involve large uncertainties caused by instrumental noise. Several methods have been proposed to solve this problem. In general, the displacement data from a nearby GPS station are used as the reference to optimize the empirical baseline correction of the accelerometer records. Tu et al. (2013) developed a method that integrates a single-frequency GPS and microelectromechanical systems (MEMS) accelerometer. In summary, our experimental results have shown that the combined system of a single-frequency GPS and a MEMS accelerometer can monitor broadband ground motion associated with strong acceleration events with precision satisfying most current demands in real-time seismology and earthquake engineering, with potential applications in other fields as well. However, dual-frequency GPS and broadband seismometer are still required and cannot be replaced by such a low-cost system.

In addition to GPS, InSAR is also widely used to understand the dynamics of an area. The accuracy of this technique is comparable with kinematic GPS, which is about a few centimeters. Unlike GPS, the InSAR technique is based on complex data measured by radar systems in the form of amplitudes and phase information. Both GPS and InSAR methods provide continuous spatiotemporal deformation data important for earthquake studies (Tralli et al., 2005). Additionally, recognizing seismic sources require information on surface deformation, frictional properties of faults, and the mechanical properties of the earth's crust and litho-sphere. Both GPS and InSAR techniques provide information that helps to produce thematic layers in GIS, representing these parameters for seismic source identification. In addition, these parameters in the form of GIS data help to determine what controls the spatial and temporal characteristics of earthquakes (Tralli et al., 2005).

11.4.3.1 *SEISMIC HAZARD AND RISK ASSESSMENT*

An important initial distinction to make in seismic hazard analysis is the difference between hazard and risk. Although these terms may seem synonymous with colloquial usage, they are not. A proper distinction must be made to facilitate correlatable discussions.

Seismic hazard refers to the actual physical result of an earthquake. Two useful, concise definitions of seismic hazard come from Reiter (1990): "the potential for dangerous, earthquake-related natural phenomena such as ground shaking, fault rupture, or soil liquefaction" and from McGuire (2004): "a property of an earthquake that can cause damage and loss" (both definitions are quoted by Wang, 2011). A seismic hazard can be completely described by the following characteristics (Wang, 2011):

- Severity (some measurement of magnitude),
- Where it occurred (some measurement in space), and
- When it occurred (some measurement in time).

On the other hand, risk describes the probability of the effects of a hazard. Again, two useful definitions come from Reiter (1990): "the probability of occurrence of these consequences (i.e., adverse consequences to society such as the destruction of buildings or the loss of life that could result from seismic hazards)" and from McGuire (2004): "the probability that some humans will incur loss or that their built environment will be damaged" (both definitions are quoted by Wang, 2011). Seismic risk is defined by the same three parameters as a seismic hazard and an additional fourth parameter: probability. Risk can be characterized by the intersection of hazard and vulnerability (ICEF, 2011; Wang, 2011). Although an area is seismically active and the hazard of ground shaking exists, if the infrastructure has been engineered in such a way that it can withstand the shaking, the seismic risk may be low compared with an area with fewer infrastructural preparations. Analysis of risk requires the use of an appropriate statistical model, commonly Poisson, empirical, Brownian passage time, or time predictable models (Wang, 2011).

11.4.3.2 DAMAGE ASSESSMENT

Earthquake hazards are not limited to the immediate event, but damage to structures can further endanger the public and emergency responders because of structural instability intensified by aftershocks (Bialas et al., 2016). Furthermore, damage to transportation and other utilities can hamper response by emergency responders and evacuation of the public from affected areas (Coburn and Spence, 2002). Thus, developing tools for damage assessment is important for emergency response to an earthquake.

Many earth observation systems can be used in detecting earthquake damage, but such systems have both advantages and drawbacks. Passive systems detect emitted or reflected spectral data, which measure the reflected signal sent by the system itself, from ground targets and active sensors. One major advantage of passive systems is that they exist at all levels of spectral and spatial resolutions, making them suitable for many applications including earthquake damage detection. Passive systems also have the capability to capture 3D information when stereo pairs of the same scene are acquired. On the other hand, active sensors perform data acquisition in all weather conditions, day and night, in addition to their availability at various spatial resolutions. They also provide 3D information similar to InSAR technique (Bossler et al., 2004). However, active sensors lack spectral information. Both passive and active sensors can be mounted on ground, aerial, and space platforms. For earthquake damage assessment, these sensors are preferred to be mounted on space and aerial platforms instead of on the ground to allow fast mapping over large areas. A comparison of space platforms with aerial ones shows that the former can easily visit remote areas regularly. The higher altitude also permits observing larger areas in a shorter time. However, the high altitude of space platforms generates spatial resolution challenges. Additionally, temporal resolution is another important characteristic of passive and active systems that play an important role in earthquake damage assessment. Satellite images before and after earthquakes must be provided for reference in damage assessment and loss estimation. The revisit time of sensors on space platforms is lengthy, and shorter revisits can be achieved with aerial platforms. Some systems can operate off-nadir to increase the temporal resolution of the system. However, additional issues arise in contending with the resultant shadows and other factors from oblique imagery. Finally, space-based systems have additional challenges owing to atmospheric conditions; weather and dust can obscure the views on the ground similar to passive systems. On the other hand, aerial platforms have several advantages over other platforms including high temporal and spatial resolution and system flexibility. Other advantages of these systems are a better capability in dealing with atmospheric conditions and potential use for nonimaging sensors such as LiDAR.

In addition, other data forms such as historical topography, cadastral maps, GIS data, government reports, and data gathered from interviews with the affected people can also help improve the quality of earthquake damage assessment. Geospatial data such as freely available roads, buildings, and other natural and man-made features assist in the classification and change detection process. Other types of geospatial data such as the Internet and social media information are also useful for geospatially significant, timely information on natural disasters that might aid in earthquake damage assessment (Yin et al., 2012).

Damage assessment is usually done by change detection between two or more sequential satellite images or aerial photographs. The images are first classified via per-pixel (Khatami et al., 2016) or object-based image classification (Sun and Vu, 2016; Bai et al., 2017) into several land cover and land use classes. Then, the results from classification and postprocessing are compared to calculate the areas that have been changed during the period between the two images acquired. The results of change detection provide both qualitative and quantitative assessments of the damaged and other changed areas. Some studies also quantify the damages into costs for detailed information on the areas affected by earthquakes.

The per-pixel classification methods usually produce outputs with salt-and-paper noises, especially in very high-resolution images such as those captured by drones, which affect the quantitative assessment of the damage estimation. In object-based methods, the results are often less noisy, data are GIS ready, and additional geometric attributes can be easily calculated for the image objects. By contrast, pixel-based methods are not open to straightforward analysis. An additional advantage of object-based classification is the removal of misclassified objects by advanced texture and contextual features. Therefore, the better result in damage assessment can be acquired through object-based classification.

11.5   WAY FORWARD

Mitigation includes all the activities and measures installed to minimize the effects and impacts of the disaster. This condition covers physical, economic, and social effects of the disaster on a given population. History has shown that adequate preparation for disasters significantly mitigates their adverse effects. Many aspects and angles are possible depending on the locality and the community concerned. Mitigation measures can be categorized as structural and nonstructural. Preparation initiatives of the government and local authorities must have clear structures for handling disasters when they occur. Simple and easy to implement plans and policies are vital for a good disaster preparedness and management unit. Educating the public and members of organizations is required, warning systems should be in place and serviced regularly, and land use should be well planned. Poor planning results in uncontrollable disasters.

GIS is one of the powerful tools that can be used for the assessment of risks of many natural hazards. GIS techniques make hazard mapping and risk assessments of various natural disasters possible in many areas around the world. GIS maps can help authorities conduct a quick and more effective assessment of the potential impact of disasters and initiation of appropriate measures to reduce the impact. GIS data can help planners and decision-makers take positive and timely steps. GIS applications should consider long-term integration, including the vertical integration that involves different application (and potential) levels and horizontal integration that involves other interest groups. Therefore, issues must be addressed from database design and data sharing to tool making (analysis functions) and experience sharing.

GPS data provide information to monitor several natural hazards such as earthquakes and landslides in real time. The accuracy of GPS measurements and their online streaming are the most important advantages in designing early warning systems for the mentioned natural disasters. The reviewed studies indicated that the use of GPS data in such applications, individually or in an integrated framework with SAR and seismic sensors, could provide an effective tool to build real-time systems that serve many citizens in a particular region. However, several limitations are found for the GPS systems in natural hazard monitoring. First, the cost of GPS receiver and their correction services hinder the placement of a dense GPS network that can accurately measure the displacement of earthquakes or other natural hazards. Second, the design of an advanced method for analyzing GPS data and frameworks that best integrates GPS with other sensors is difficult and requires extensive computational efforts. The future directions for GPS use for natural hazard monitoring should focus on designing accurate GPS systems with low cost and developing robust algorithms that can process GPS data in real time to accurately calculate the natural hazard parameters in the field and predict a risk event.

Satellite and aerial images (optical and radar) are great sources of data for natural disaster management. The use of these data has been proven helpful for various applications including landslide detection, monitoring, and modeling, oil spill detection and trajectory simulations, earthquake studies, and damage assessment of the areas affected by natural disasters. Their main advantages include images at various spatial and temporal resolutions. However, acquiring image data in a rapid response situation remains a challenge, both technically and financially. Optical data offer several advantages over SAR but are inherently affected by cloud cover, smoke, or haze at the time of satellite overpass. The flexibility provided by a multisensory, multiplatform approach is likely to provide the most comprehensive coverage of a disaster event.

LiDAR data are effective for landslide and earthquake studies. The very high-resolution DEM and digital surface models provide accurate 3D information for detailed damage assessment and landslide detection, susceptibility mapping, and risk assessments. In addition, LiDAR data are also useful for mapping surface expressions of faulting, which can be useful for deformation mapping. However, the main challenge of LiDAR data is its processing cost. For hazard mapping and monitoring of rapid natural disasters, such as landslide and earthquake, processing LiDAR data can take a long time. Thus, it is not appropriate for providing informa- tion in a rapid response emergency where it can be manually interpreted.

Although geospatial technology has advanced the applications of the landslide, oil spill, and earthquake, several challenges still remain. First, the use of LiDAR data for landslide detection and spatial prediction requires robust algorithms for filtering the data. LiDAR data provide detailed infor- mation on the structural and geomorphological features; however, some of the information is lost during filtering to DEM. With more advanced filtering techniques, these features can be preserved and better landslide recognition and prediction can be achieved. Second, oil spill detection and trajectory simulations require multisource datasets that may be expensive. Detection of the oil spill is better achieved by SAR images, its thickness could be estimated by microwave radiometer, and the type of oil could be identified by hyperspectral sensors or laser fluorescence. In addition, the simulation of oil spill trajectory requires information on wind and ocean condition, which may not be available at fine scales. Methods designed to study oil spill detection and modeling with a single sensor can be a great advancement, with the potential to save processing time and significantly reduce the cost. Third, monitoring large areas where earth shaking is expected can be expensive with current GPS units and technologies. In the future, cheaper GPS units will be necessary so that large and dense GPS networks can be placed in these areas for better monitoring and prepara- tion for disasters such as earthquakes.

11.6 CONCLUSION

This chapter presented the role of geospatial information for advancing landslide, oil spill, and earthquake applications. The geospatial technologies such as GIS, GPS, SAR, and LiDAR are briefly explained after introducing the four-part cycle of natural disaster management. Then, the landslide, oil spill, and earthquake applications are presented and discussed in detail. Finally, the limitations and challenges of using geospatial data to assess and model natural disasters are presented. The future directions of geospatial technology for natural disaster management must focus on developing GIS functions that can store complex data from multiple sources more efficiently and process spatiotemporal data more effectively in a shorter time. GPS units must be designed for lower costs, and their integration with other seismic sensors should be improved. Designing systems with multisensory approach can provide improved solutions for monitoring natural disasters. Finally, enhancing the filtering algorithms of LiDAR point clouds can be a significant contribution to advancing landslide and earthquake applications.

KEYWORDS

- **GIS**
- **remote sensing**
- **natural hazards**
- **mapping**
- **monitoring**

REFERENCES

Abdulwahid, W. M.; Pradhan, B. Landslide Vulnerability and Risk Assessment for Multi-hazard Scenarios Using Airborne Laser Scanning Data (LiDAR). *Landslides* **2017,** *14* (3), 1057–1076.

Akgun, A.; Kıncal, C.; Pradhan, B. Application of Remote Sensing Data and GIS for Landslide Risk Assessment as an Environmental Threat to Izmir City (West Turkey). *Environ. Monit. Assess.* **2012,** *184* (9), 5453–5470.

Akgun, A.; Erkan, O. Landslide Susceptibility Mapping by Geographical Information System-based Multivariate Statistical and Deterministic Models: In an Artificial Reservoir Area at Northern Turkey. *Arab. J. Geosci.* **2016,** *9* (2), 1–15.

Althuwaynee, O. F.; Pradhan, B. Semi-quantitative Landslide Risk Assessment Using GIS-based Exposure Analysis in Kuala Lumpur City. *Geomat. Nat. Haz. Risk* **2017,** *8,* 706–732.

Bai, Y.; Adriano, B.; Mas, E.; Gokon, H.; Koshimura, S. Object-based Building Damage Assessment Methodology Using Only Post Event ALOS-2/PALSAR-2 Dual Polarimetric SAR Intensity Images. *J. Disaster Res.* **2017,** *12* (2), 259–271.

Bellone, T.; Dabove, P.; Manzino, A. M.; Taglioretti, C. Real-time Monitoring for Fast Deformations Using GNSS Low-cost Receivers. *Geomat. Nat. Haz. Risk* **2016**, *7* (2), 458–470.

Bialas, J.; Oommen, T.; Rebbapragada, U.; Levin, E. Object-based Classification of Earthquake Damage from High-resolution Optical Imagery Using Machine Learning. *J. Appl. Remote Sens.* **2016**, *10* (3), 036025–036025.

Biswajeet, P.; Hamid, A. Oil Spill Trajectory Simulation and Coastal Sensitivity Risk Mapping. *Res. J. Chem. Environ.* **2009**, *13* (4), 73–80.

Bossler, J. D.; Campbell, J. B.; Mcmaster, R. B.; Rizos, C.; Eds. *Manual of Geospatial Science and Technology*; CRC Press: Boca Raton, 2010.

Bovenga, F.; Pasquariello, G.; Pellicani, R.; Refice, A.; Spilotro, G. Landslide Monitoring for Risk Mitigation by Using Corner Reflector and Satellite SAR Interferometry: The Large Landslide of Carlantino (Italy). *Catena* **2017**, *151*, 49–62.

Brekke, C.; Solberg, A. H. Oil Spill Detection by Satellite Remote Sensing. *Remote Sens. Environ.* **2005**, *95* (1), 1–13.

Brown, C. E.; Fingas, M. F. Review of the Development of Laser Fluorosensors for Oil Spill Application. *Mar. Pollut. Bull.* **2003**, *47* (9), 477–484.

Bui, D. T.; Pradhan, B.; Revhaug, I.; Nguyen, D. B.; Pham, H. V.; Bui, Q. N. A Novel Hybrid Evidential Belief Function-based Fuzzy Logic Model in Spatial Prediction of Rainfall-induced Shallow Landslides in the Lang Son City Area (Vietnam). *Geomat. Nat. Haz. Risk* **2015**, *6* (3), 243–271.

Byfield, V.; Boxall, S. R. In *Thickness Estimates and Classification of Surface Oil Using Passive Sensing at Visible and Near-infrared Wavelengths*, Proceedings of the IEEE 1999 International Geoscience and Remote Sensing Symposium (IGARSS'99), 1999; IEEE, 1999; Vol. 3, pp 1475–1477.

Chen, K.; Babeyko, A.; Hoechner, A.; Ge, M. Comparing Source Inversion Techniques for GPS-based Local Tsunami Forecasting: A Case Study for the April 2014 M8.1 Iquique, Chile, Earthquake. *Geophys. Res. Lett.* **2016**, *43* (7), 3186–3192.

Cheng, G.; Guo, L.; Zhao, T.; Han, J.; Li, H.; Fang, J. Automatic Landslide Detection from Remote-sensing Imagery Using a Scene Classification Method Based on BoVW and pLSA. *Int. J. Remote Sens.* **2013**, *34* (1), 45–59.

Coburn, A.; Spence, R. *Earthquake Protection*; John Wiley & Sons, 2003 (ISO 690).

Colombelli, S.; Allen, R. M.; Zollo, A. Application of Real-time GPS to Earthquake Early Warning in Subduction and Strike-slip Environments. *J. Geophys. Res.* **2013**, *118* (7), 3448–3461.

Corominas, J.; Van Westen, C.; Frattini, P.; Cascini, L.; Malet, J. P.; Fotopoulou, S.; Pitilakis, K. Recommendations for the Quantitative Analysis of Landslide Risk. *Bull. Eng. Geol. Environ.* **2014**, *73* (2), 209–263.

De Paratesi, S.; Barrett, E. Hazards and Disasters: Concepts and Challenges. In *Remote Sensing for Hazard Monitoring and Disaster Assessment: Marine and Coastal Applications in the Mediterranean Region*; CRC Press, Gordon and Breach Science Publishers: Singapore, 1989; pp 1–17.

Devkota, K. C.; Regmi, A. D.; Pourghasemi, H. R.; Yoshida, K.; Pradhan, B.; Ryu, I. C.; Dhital, M. R.; Althuwaynee, O. F. Landslide Susceptibility Mapping Using Certainty Factor, Index of Entropy and LR Models in GIS and Their Comparison at Mugling–Narayanghat Road Section in Nepal Himalaya. *Nat. Hazards* **2013**, *65* (1), 135–165.

Domínguez-Cuesta, M. J.; Jiménez-Sánchez, M.; Berrezueta, E. Landslides in the Central Coalfield (Cantabrian Mountains, NW Spain): Geomorphological Features, Conditioning Factors and Methodological Implications in Susceptibility Assessment. *Geomorphology* **2007**, *89* (3), 358–369.

Ferraro, G.; Meyer-Roux, S.; Muellenhoff, O.; Pavliha, M.; Svetak, J.; Tarchi, D.; Topouzelis, K. Long Term Monitoring Of Oil Spills in European Seas. *Int. J. Remote Sens.* **2009**, *30* (3), 627–645.

Fingas, M.; Brown, C. An Update on Oil Spill Remote Sensors. In Proceedings of the 28 Arctic and Marine Oilspill Program (AMOP) Technical Seminar, (p. 1134). Canada: Environment Canada, 2005.

Fuchs, S.; Heiss, K.; Hübl, J. Towards an Empirical Vulnerability Function for Use in Debris Flow Risk Assessment. *Nat. Hazards Earth Syst. Sci.* **2007**, *7* (5), 495–506.

Garcia-Pineda, O.; MacDonald, I.; Hu, C.; Svejkovsky, J.; Hess, M.; Dukhovskoy, D.; Morey, S. L. Detection of Floating Oil Anomalies from the Deepwater Horizon Oil Spill with Synthetic Aperture Radar. *Oceanography* **2013**, *26* (2), 124–137.

Goodman, R. Overview and Future Trends in Oil Spill Remote Sensing. *Spill Sci. Technol. Bull.* **1994**, *1* (1), 11–21.

Gorsevski, P. V.; Brown, M. K.; Panter, K.; Onasch, C. M.; Simic, A.; Snyder, J. Landslide Detection and Susceptibility Mapping Using LiDAR and an Artificial Neural Network Approach: A Case Study in the Cuyahoga Valley National Park, Ohio. *Landslides* **2016**, *13* (3), 467–484.

Harp, E. L.; Keefer, D. K.; Sato, H. P.; Yagi, H. Landslide Inventories: The Essential Part of Seismic Landslide Hazard Analyses. *Eng. Geol.* **2011**, *122* (1), 9–21.

Hong, H.; Chen, W.; Xu, C.; Youssef, A. M.; Pradhan, B.; Tien Bui, D. Rainfall-induced Landslide Susceptibility Assessment at the Chongren Area (China) Using Frequency Ratio, Certainty Factor, and Index Of Entropy. *Geocarto Int.* **2017a**, *32* (2), 139–154.

Hong, H.; Pradhan, B.; Bui, D. T.; Xu, C.; Youssef, A. M.; Chen, W. Comparison of Four Kernel Functions Used in Support Vector Machines for Landslide Susceptibility Mapping: A Case Study at Suichuan Area (China). *Geomat. Nat. Haz. Risk* **2017b**, *8*, 544–569.

Janalipour, M.; Mohammadzadeh, A. Building Damage Detection Using Object-based Image Analysis and ANFIS from High-resolution Image (Case Study: BAM Earthquake, Iran). *IEEE J. Sel. Top. Appl. Earth Obs. Remote Sens.* **2016**, *9* (5), 1937–1945.

Jebur, M. N.; Pradhan, B.; Tehrany, M. S. Detection of Vertical Slope Movement in Highly Vegetated Tropical Area of Gunung Pass Landslide, Malaysia, Using L-band InSAR Technique. *Geosci. J.* **2014a**, *18* (1), 61–68.

Jebur, M. N.; Pradhan, B.; Tehrany, M. S. Optimization of Landslide Conditioning Factors Using Very High-resolution Airborne Laser Scanning (LiDAR) Data at Catchment Scale. *Remote Sens. Environ.* **2014b**, *152*, 150–165.

Jha, M. N.; Levy, J.; Gao, Y. Advances in Remote Sensing for Oil Spill Disaster Management: State-of-the-art Sensors Technology for Oil Spill Surveillance. *Sensors* **2008**, *8* (1), 236–255.

Khatami, R.; Mountrakis, G.; Stehman, S. V. A Meta-analysis of Remote Sensing Research on Supervised Pixel-based Land-cover Image Classification Processes: General Guidelines for Practitioners and Future Research. *Remote Sens. Environ.* **2016**, *177*, 89–100.

Komac, M.; Holley, R.; Mahapatra, P.; van der Marel, H.; Bavec, M. Coupling of GPS/ GNSS and Radar Interferometric Data for a 3D Surface Displacement Monitoring Of Landslides. *Landslides* **2015**, *12* (2), 241–257.

Lacroix, P.; Zavala, B.; Berthier, E.; Audin, L. Supervised Method of Landslide Inventory Using Panchromatic SPOT5 Images and Application to the Earthquake-triggered Landslides of Pisco (Peru, 2007, Mw 8.0). *Remote Sens.* **2013**, *5* (6), 2590–2616.

Leone, F.; Asté, J. P.; Leroi, E. *Vulnerability Assessment of Elements Exposed to Mass-movement: Working Toward a Better Risk Perception.* In *Landslides-Glissements de Terrain*; Balkema: Rotterdam, 1996; pp 263–270.

Leprince, S.; Berthier, E.; Ayoub, F.; Delacourt, C.; Avouac, J. P. Monitoring Earth Surface Dynamics with Optical Imagery. *Eos Trans. AGU* **2008,** *89* (1), 1–2.

Lewis, S. *Remote Sensing for Natural Disasters: Facts and Figures*; Science and Development Network, 2009.

Li, Z.; Shi, W.; Lu, P.; Yan, L.; Wang, Q.; Miao, Z. Landslide Mapping from Aerial Photographs Using Change Detection-based Markov Random Field. *Remote Sens. Environ.* **2016,** *187*, 76–90.

Lian, X.; Hu, H. Terrestrial Laser Scanning Monitoring and Spatial Analysis of Ground Disaster in Gaoyang Coal Mine in Shanxi, China: A Technical Note. *Environ. Earth Sci.* **2017,** *76* (7), 287.

Mann, U.; Pradhan, B.; Prechtel, N.; Buchroithner, M. F. An Automated Approach for Detection of Shallow Landslides from LiDAR Derived DEM Using Geomorphological Indicators in a Tropical Forest. In *Terrigenous Mass Movements*; Springer Berlin Heidelberg, 2012; pp 1–22.

Marghany, M. RADARSAT for Oil Spill Trajectory Model. *Environ. Modell. Softw.* **2004,** *19* (5), 473–483.

Michael, E. A.; Samanta, S. Landslide Vulnerability Mapping (LVM) Using Weighted Linear Combination (WLC) Model Through Remote Sensing and GIS Techniques. *Model. Earth Syst. Environ.* **2016,** *2* (2), 1–15.

Muthukumar, M. GIS Based Geosystem Response Modeling for Landslide Vulnerability Mapping Parts of Nilgiris, South India. *Disaster Adv.* **2013,** *6* (7), 58–66.

U'loole, T. B.; Valentine, A. P.; Woodhouse, J. H. Earthquake Source Parameters from GPS-measured Static Displacements with Potential for Real-time Application. *Geophys. Res. Lett.* **2013,** *40* (1), 60–65.

Pelyushenko, S. A. Microwave Radiometer System for the Detection of Oil Slicks. *Spill Sci. Technol. Bull.* **1995,** *2* (4), 249–254.

Pham, B. T.; Bui, D. T.; Prakash, I.; Dholakia, M. B. Hybrid Integration of Multilayer Perceptron Neural Networks and Machine Learning Ensembles for Landslide Susceptibility Assessment at Himalayan Area (India) Using GIS. *Catena* **2017,** *149*, 52–63.

Polcari, M.; Palano, M.; Fernández, J.; Samsonov, S. V.; Stramondo, S.; Zerbini, S. 3D Displacement Field Retrieved by Integrating Sentinel-1 InSAR and GPS Data: The 2014 South Napa Earthquake. *Eur. J. Remote Sens.* **2016,** *49* (1), 1–13.

Pourghasemi, H. R.; Beheshtirad, M.; Pradhan, B. A Comparative Assessment of Prediction Capabilities of Modified Analytical Hierarchy Process (M-AHP) and Mamdani Fuzzy Logic Models Using Netcad-GIS for Forest Fire Susceptibility Mapping. *Geomat. Nat. Haz. Risk* **2016,** *7* (2), 861–885.

Pradhan, B.; Lee, S.; Buchroithner, M. F. Use of Geospatial Data and Fuzzy Algebraic Operators to Landslide-hazard Mapping. *Appl. Geomat.* **2009a,** *1* (1–2), 3–15.

Pradhan, B.; Shafiee, M.; Pirasteh, S. Maximum flood Prone Area Mapping Using RADARSAT Images and GIS: Kelantan River Basin. *Int. J. Geoinform.* **2009b,** *5* (2), 11.

Pradhan, B.; Hagemann, U.; Tehrany, M. S.; Prechtel, N. An Easy to Use ArcMap Based Texture Analysis Program for Extraction of Flooded Areas from TerraSAR-X Satellite Image. *Comput. Geosci.* **2014,** *63*, 34–43.

Pradhan, B.; Jebur, M. N.; Shafri, H. Z. M.; Tehrany, M. S. Data Fusion Technique Using Wavelet Transform and Taguchi Methods for Automatic Landslide Detection from Airborne Laser Scanning Data and Quickbird Satellite Imagery. *IEEE Trans. Geosci. Remote Sens.* **2016,** *54* (3), 1610–1622.

Price, J. M.; Reed, M.; Howard, M. K.; Johnson, W. R.; Ji, Z. G.; Marshall, C. F.; Norman, L. G.; Rainey, G. B. Preliminary Assessment of an Oil-spill Trajectory Model Using Satellite-tracked, Oil-spill-simulating Drifters. *Environ. Modell. Softw.* **2006,** *21* (2), 258–270.

Raja, N. B.; Çiçek, I.; Türkoğlu, N.; Aydin, O.; Kawasaki, A. Landslide Susceptibility Mapping of the Sera River Basin Using Logistic Regression Model. *Nat. Hazards* **2017,** *85* (3), 1323–1346.

Riquelme, S.; Bravo, F.; Melgar, D.; Benavente, R.; Geng, J.; Barrientos, S.; Campos, J. W Phase Source Inversion Using High-rate Regional GPS Data for Large Earthquakes. *Geophys. Res. Lett.* **2016,** *43* (7), 3178–3185.

Romeo, R. W.; Mari, M.; Pappafico, G. A Performance-based Approach to Landslide Risk Analysis and Management. In *Landslide Science and Practice*; Springer Berlin Heidelberg, 2013; pp 91–95.

Rossi, G.; Nocentini, M.; Lombardi, L.; Vannocci, P.; Tanteri, L.; Dotta, G.; et al. Integration of Multicopter Drone Measurements and Ground-based Data for Landslide Monitoring. In *Landslides and Engineered Slopes. Experience, Theory and Practice*; Associazione Geotecnica Italiana: Rome, 2016; pp 1745–1750.

Salem, F. M. F. *Hyperspectral Remote Sensing: A New Approach for Oil Spill Detection and Analysis*; 2003.

Sangchini, E. K.; Emami, S. N.; Tahmasebipour, N.; Pourghasemi, H. R.; Naghibi, S. A.; Arami, S. A.; Pradhan, B. Assessment and Comparison of Combined Bivariate and AHP Models with Logistic Regression for Landslide Susceptibility Mapping in the Chaharmahal-e-Bakhtiari Province, Iran. *Arab. J. Geosci.* **2016,** *9* (3), 1–15.

Scaioni, M.; Longoni, L.; Melillo, V.; Papini, M. Remote Sensing for Landslide Investigations: An Overview of Recent Achievements and Perspectives. *Remote Sens.* **2014,** *6* (10), 9600–9652.

Sharma, R. C.; Tateishi, R.; Hara, K.; Nguyen, H. T.; Gharechelou, S.; Nguyen, L. V. Earthquake Damage Visualization (EDV) Technique for the Rapid Detection of Earthquake-induced Damages Using SAR Data. *Sensors* **2017,** *17* (2), 235.

Siyahghalati, S.; Saraf, A. K.; Pradhan, B.; Jebur, M. N.; Tehrany, M. S. Rule-based Semi-automated Approach for the Detection of Landslides Induced by 18 September 2011 Sikkim, Himalaya, Earthquake Using IRS LISS3 Satellite Images. *Geomat. Nat. Haz. Risk* **2016,** *7* (1), 326–344.

Staley, D. M.; Negri, J. A.; Kean, J. W.; Tillery, A. C.; Youberg, A. M. *Updated Logistic Regression Equations for the Calculation of Post-fire Debris-flow Likelihood in the Western United States*; US Geological Survey Open-File Report, 2016; Vol. 1106, p 13.

Strozzi, T.; Ambrosi, C.; Raetzo, H. Interpretation of Aerial Photographs and Satellite SAR Interferometry for the Inventory of Landslides. *Remote Sens.* **2013,** *5* (5), 2554–2570.

Sun, J.; Vu, T. T. Distributed and Hierarchical Object-based Image Analysis for Damage Assessment: A Case Study of 2008 Wenchuan Earthquake, China. *Geomat. Nat. Haz. Risk* **2016,** *7* (6), 1962–1972.

Sun, B.; Xu, Q.; He, J.; Liu, Z.; Wang, Y.; Ge, F. Damage Assessment Framework for Landslide Disaster Based on Very High-resolution Images. *J. Appl. Remote Sens.* **2016,** *10* (2), 025027–025027.

Tien Bui, D.; Tuan, T. A.; Hoang, N. D.; Thanh, N. Q.; Nguyen, D. B.; Van Liem, N.; Pradhan, B. Spatial Prediction of Rainfall-induced Landslides for the Lao Cai Area (Vietnam) Using a Hybrid Intelligent Approach of Least Squares Support Vector Machines Inference Model and Artificial Bee Colony Optimization. *Landslides* **2017,** *14* (2), 447–458.

Tralli, D. M.; Blom, R. G.; Zlotnicki, V.; Donnellan, A.; Evans, D. L. Satellite Remote Sensing of Earthquake, Volcano, Flood, Landslide and Coastal Inundation Hazards. *ISPRS J. Photogramm. Remote Sens.* **2005,** *59* (4), 185–198.

Tu, R.; Wang, R.; Ge, M.; Walter, T. R.; Ramatschi, M.; Milkereit, C.; Bindi, D.; Dahm, T. Cost-effective Monitoring of Ground Motion Related to Earthquakes, Landslides, or Volcanic Activity by Joint Use of a Single-frequency GPS and a MEMS Accelerometer. *Geophys. Res. Lett.* **2013,** *40* (15), 3825–3829.

Turker, M.; Cetinkaya, B. Automatic Detection of Earthquake-damaged Buildings Using DEMs Created from Pre- and Post-earthquake Stereo Aerial Photographs. *Int. J. Remote Sens.* **2005,** *26* (4), 823–832.

Ural, S.; Hussain, E.; Kim, K.; Fu, C. S.; Shan, J. Building Extraction and Rubble Mapping for City Port-au-Prince Post-2010 Earthquake with GeoEye-1 Imagery and Lidar Data. *Photogramm. Eng. Remote Sens.* **2011,** *77* (10), 1011–1023.

Van der Sande, C. J.; De Jong, S. M.; De Roo, A. P. J. A Segmentation and Classification Approach of IKONOS-2 Imagery for Land Cover Mapping to Assist Flood Risk and Flood Damage Assessment. *Int. J. Appl. Earth Observ. Geoinform.* **2003,** *4* (3), 217–229.

Van Westen, C. J.; Soeters, R. In *Remote Sensing and Geographic Information Systems for Natural Disaster Management*, Natural Disasters and Their Mitigation: A Remote Sensing and GIS Perspective; Roy, P., van Westen, C. J., Ray P. C., Eds.; Indian Institute of Remote Sensing, National Remote Sensing Agency, India, 2000; pp 31–76.

Van Westen, C. J.; Getahun, F. L. Analyzing the Evolution of the Tessina Landslide Using Aerial Photographs and Digital Elevation Models. *Geomorphology* **2003,** *54* (1), 77–89.

Van Westen, C. J. Geo-information Tools for Landslide Risk Assessment: An Overview of Recent Developments. In *Landslides: Evaluation and Stabilization*; Balkema, Taylor & Francis Group: London, 2004; Vol. 1, pp 39–56.

Varnes, D. J. *Landslide Hazard Zonation: A Review of Principles and Practice;* No. 3, 1984.

Vassilopoulou, S.; Hurni, L.; Dietrich, V.; Baltsavias, E.; Pateraki, M.; Lagios, E.; Parcharidis, I. Orthophoto Generation Using IKONOS Imagery and High-resolution DEM: A Case Study on Volcanic Hazard Monitoring of Nisyros Island (Greece). *ISPRS J. Photogramm. Remote Sens.* **2002,** *57* (1), 24–38.

Vigny, C.; Simons, W. J. F.; Abu, S.; Bamphenyu, R.; Satirapod, C.; Choosakul, N.; Ambrosius, B. A. C. Insight into the 2004 Sumatra–Andaman Earthquake from GPS Measurements in Southeast Asia. *Nature* **2005,** *436* (7048), 201–206.

Williams, K.; Olsen, M. J.; Roe, G. V.; Glennie, C. Synthesis of Transportation Applications of Mobile LiDAR. *Remote Sens.* **2013,** *5* (9), 4652–4692.

Winter, M. G.; Smith, J. T.; Fotopoulou, S.; Pitilakis, K.; Mavrouli, O.; Corominas, J.; Argyroudis, S. An Expert Judgement Approach to Determining the Physical Vulnerability of Roads to Debris Flow. *Bull. Eng. Geol. Environ.* **2014,** *73* (2), 291–305.

Wübbena, G.; Bagge, A.; Boettcher, G.; Schmitz, M.; Andree, P. In *Permanent Object Monitoring with GPS with 1 Millimeter Accuracy*, International Technical Meeting ION GPS-01, Salt Lake City, USA, September, 2001.

Xie, M. W.; Jia, Y. C.; Lv, F. X.; He, B.; Chang, S. X. A Comprehensive Information System for Landslide Monitoring Based on a Three-dimensional Geographic Information System. In *Green Building, Environment, Energy and Civil Engineering: Proceedings of the 2016 International Conference on Green Building, Materials and Civil Engineering (GBMCE 2016), April 26–27 2016, Hong Kong, PR China*; CRC Press, 2016; p 289.

Yin, J.; Lampert, A.; Cameron, M.; Robinson, B.; Power, R. Using Social Media to Enhance Emergency Situation Awareness. *IEEE Intell. Syst.* **2012,** *27* (6), 52–59.

CHAPTER 12

Environmental Impacts of Renewable Energy

NAUSHEEN MAZHAR* and SAHAR ZIA

Department of Geography, Lahore College for Women University, Lahore, Pakistan

Corresponding author. E-mail: noshmazhar@gmail.com

ABSTRACT

This chapter summarizes the significance of renewable energy resources, their technological practices, positive and negative impacts. It also presents current energy scenario and future prospects. Projected demand of energy will increase drastically by 2030 as per findings of International Energy Agency (IEA). Renewable energy sources must be used at a massive scale, as a supplement to conventional energy sources, in order to fulfill the ever-increasing demand of energy. However, renewable forms of energy like solar, wind, hydropower, ocean, geothermal, and biofuels are not completely environment friendly, since they do have their impacts on biodiversity, soil, aquatic environments, marine life, air quality, and even human health. However, in comparison to the conventional sources of energy, their negative impacts on environment are quite meager, in most cases these sources of energy production proved to be carbon neutral and in some cases carbon negative. The chapter concludes that reliance on renewable forms of energy for power generation, will become indispensable for a sustainable future.

12.1 INTRODUCTION

Energy is one of the basic inputs for the socioeconomic development of any country. The International Energy Agency (IEA) projected that since 1990, there has been a 40% increase in the demand for energy worldwide and this demand is expected to rise to 53% by 2030. Energy access is not uniform across the globe as almost 2 billion people still do not have access to electricity. There exists a great misbalance between the population and its energy consumption, as the industrialized nations of the world consume relatively a huge share of the world's total energy, in comparison to their small population, for example, the United States is a home to 5% of the world's population but consumes 20% of the world's total energy. Botkin and Keller (2012) mentioned the grave issue of energy shortfalls in future in their book by stating that there would be a great uncertainty in terms of "energy availability and cost" in future. Economic factors usually surpass the environmental concerns, while the choices of energy are being made. However, while the world economy is dependent to a great extent on fuel prices, their scarcity and depletion are nonetheless becoming a grave issue.

Besides energy shortfalls, the preference of fossil fuels for generating electricity worldwide is causing changes in the climate through the green-house gas (GHG) emissions. Environmental Protection Agency (EPA) concludes that in the United States alone 67% of the electricity is produced through the burning of fossil fuels, which was 29% of the total GHG emissions, in 2015. Use of fossil fuel as a source of power generation also leads to the addition of nitrogen oxides (NO_x) and methane (CH_4). The concentration of carbon dioxide in the atmosphere has increased to 30%, since the last 30 years which led to a rise of 0.6°C of average global surface temperature since the past 100 years (Herbert and Krishnan, 2016). Elliot, 2003 affirms that 80% of the global energy still comes from the fossil fuels. According to the Royal Commission on Environmental Pollution, 2000; IPCC recommends that the concentration of carbon dioxide in the upper troposphere must be reduced to 550 ppm by volume, in order to curtail the negative and severe impacts of climate change. This 550 ppm implies a "60% reduction in GHG emissions (from 1999 levels)."

Due to such grave issues with power production, there was a shift in the energy preferences from fossil fuels to renewable sources of energy in the 1990s all over the world. The use of solar, wind, and geothermal energy got more fame than the use of fossil fuels. The world is quickly shifting toward the renewable forms of energy, as is evident by the fact presented in the REN21 Renewables 2017 Global Status Report, "renewables accounted for an estimated nearly 62% of net additions to global power generating capacity." Renewable sources of energy are "those sources that are replenished by natural processes so that they can be used indefinitely." According to Klugmann, 2014, renewable energy sources contribute 19% toward the world's total electricity production. Figure 12.1 shows the percentage of regional capacity of renewable energy. Among all, Asia shares the largest proportion with 35% of the total renewable capacity of the world.

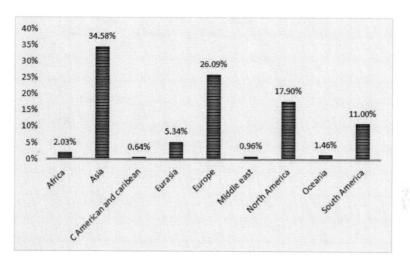

FIGURE 12.1   Regional capacity of renewable energy.

Source: Reprinted with permission from REN21, 2017.

Besides nonrenewable sources of energy, the sources of energy like nuclear, geothermal, hydropower, solar, etc., are referred to as alternate sources of energy as they offer an alternative to the conventional sources of energy like oil, gas, coal, etc. Alternate energy sources are the sources of energy that are not only renewable but also cause lesser harm to the environment than the conventional means of electricity generation. Wind, solar, besides others, are referred to as renewable sources of energy as they do not get depleted by consumption. Figure 12.2 shows that trend of solar and wind power generation has been increasing from last decade.

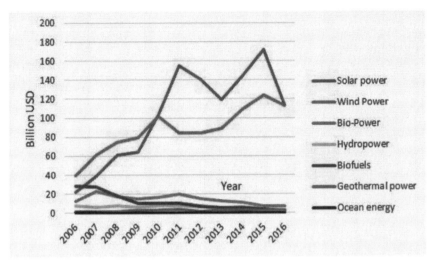

FIGURE 12.2   (See color insert.) Global investment by technology (2006–2007).

Source: Reprinted with permission from REN21, 2017.

However, hydropower capacity (GW) is still remained higher, that is, 1096 GW among all renewable resources (REN21, 2017) as shown in Figure 12.3.

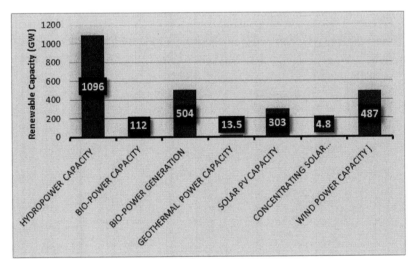

FIGURE 12.3   Renewable energy capacity by technology (GW).

Source: Reprinted with permission from REN21, 2017.

"The renewable energy resources are indigenous, non-polluting and virtually inexhaustible" mentions Herbert and Krishnan, 2016. However, some studies also speculate that renewable forms of energy also harm the environment to some extent through the wastes they produce, which might be toxic sometimes, the emissions from air and water, alterations in land use, emissions that intensify the global warming, and the noise generation. Thus these sustainable power alternate, that is, renewable forms of energy, are not all environment friendly. They cause major changes in the landscape and biodiversity of the region and involve heavy costs of installation. These sustainable sources of energy also cause air, water, soil, and noise pollution, but on a lesser scale, when compared with nonrenewable sources of energy. This chapter aims to discuss the various positive and negative impacts that the renewable forms of energy can have on the environment and to create awareness that efforts must be made to counter the environmental impacts of these forms of energy to make them all the more preferable means of power generation for a sustainable future.

12.2 HYDROPOWER

Hydropower, also called hydroelectric power, is the process of using moving water energy to create electricity as shown in Figure 12.4. It includes river flows, wave energy, and tidal energy. In this chapter, the energy produced

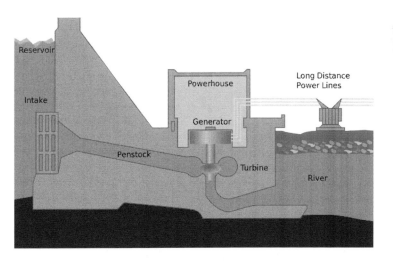

FIGURE 12.4 Schematics of hydropower plant.

Source: Tomia. (https://commons.wikimedia.org/wiki/File:Hydroelectric_dam.svg)

by running water and ocean energy shall be discussed under this heading. It is not a new technology to produce electricity as people have had knowledge about the energy potential from water for thousands of years, in fact, it is an age-old tool. Hydropower results in interactive effects, spatial and temporal on the environment, ecology, and social aspects due to reservoir inundation, flow variations over time, and river fragmentation.

Figure 12.5 presents the regional distribution of hydropower energy capacity, presenting Asia with the highest share of 41% (REN21, 2017). Australia utilizes most of the potential of its hydropower resource. It has more than 100 operating hydroelectric power stations with a total installed capacity of about 7800 megawatts (MW). These are located in the areas of highest rainfall and elevation and are mostly in New South Wales 55% and Tasmania 29% (Australia, 2010).

12.2.1   IMPACTS OF HYDROPOWER

12.2.1.1   POSITIVE IMPACTS

This mere substitution of conventional power plants based electricity generation seems to have marked environmental advantage. Although

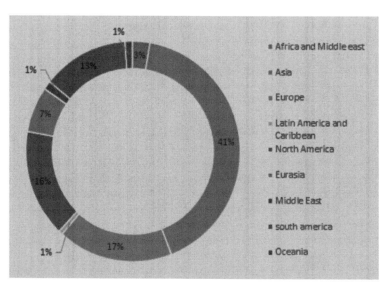

FIGURE 12.5   (See color insert.) Regional distribution of hydropower energy capacity in megawatts.

Source: Reprinted with permission from REN21, 2017.

hydropower requires relatively high initial investment, it has a long lifespan as identified by Giesecke and Mosonyi (2005) as shown in Table 12.1. Moreover, operating labor cost is also low afterward even less than $0.01/kWh (Gielen, 2012). Hydropower is remarkable for valuable energy services among other renewable energy resources, as the generating units can be started or stopped almost instantly. Due to trapping of pollutants in the form of sediments in the reservoir, water quality improves in the downstream river system.

12.2.1.2 NEGATIVE IMPACTS

Hydropower is not only environmentally friendly but also price competitive as water is free. Although it is considered as nonpolluting resource, it still is capable of changing the environment by leaving its negative impacts. For instance, during the time of less rainfall, continuity of supply of electricity through hydropower can be interrupted. Due to stagnant water, the most striking example is the increase in evaporation rates and in the air humidity of the site, especially in the case of large reservoirs. In semiarid and arid regions these processes may cause a drastic augmentation in water salinity. Reservoirs can also become problematic as these are considered as breeding grounds for disease vectors. The reservoir will have higher amounts of sediments and nutrients, which can lead to cultivating an excess of algae and other aquatic weeds.

Wildlife impact can be direct and indirect. Dam directly restricts the fish movement, upstream or downstream. In passing through the turbines,

TABLE 12.1 Lifespan of Hydropower Equipment.

Equipment	Typical lifespan of new components (years)	Equipment	Typical lifespan of new components (years)
Shut down valve	60	Cables	50
Turbine stationary parts	50	Generator protection	20
Turbine governor	30	Voltage regulator	20
Generator	40	Synchronization	20
Auxiliary transformer	40	Unit control system	20
Battery systems	20	Plant control system	20
Battery chargers	25	Emergency diesel	50

Source: Modified from Giesecke and Mosonyi, 2005.

spillways or in the diversion, fish are subjected to injury by physical contact, pressure change, shear force, or eddies. Besides direct impacts, this water is low in dissolved oxygen. Reduced oxygen and high-temperature conditions harm fish communities.

Running water is transformed into lakes while constructing dams, thus, it results in reduced water flow. Neither reservoir water is flowing like a normal river water nor have normal temperature. The water of a reservoir is also warmer in the winter and cooler in the summer. As a result, sediments accumulate and carbon, nutrients, and toxic compounds settle at the reservoir bottom.

Although GHGs emission from the reservoir is uncertain but these are also subject to concern all over the world. GHGs can be emitted in three stages: (1) construction; (2) operation and maintenance; and (3) dismantling. At the time of construction, GHG is produced by transportation of construction materials (e.g., concrete, steel, etc.) and diesel-based equipment.

GHGs are expected to emit when upper water column of the reservoir is experiencing high temperature and this high temperature helps to develop stratified conditions. For example, CO_2 is produced in oxic and anoxic conditions in the water column, and sediments of the reservoir. CH_4 is also produced under anaerobic conditions in the sediments. In some cases, NO_2 has also been found. However, it has also been noted that those areas where deforestation is not performed before dam construction are emitting more gases than emissions from conventional power plants. It has also been noted that due to poor construction material or terrorist act, catastrophic impacts can be experienced by the downriver settlements and infrastructure. However, these negative impacts can be mitigated through new sustainability guidelines, innovative planning based on stakeholder consultations, and scientific know-how.

12.3   ENERGY FROM OCEAN

The oceans cover 71% of the earth's surface and contain 97% of the earth's water. Oceans possess vast energy potential in various forms, namely, ocean wave, tidal range, tidal current, ocean current, ocean thermal energy, and salinity gradient. The world's oceans serve as a huge reserve of renewable energy; however, ocean energy is not devoid of its potential to harm the environment during its installation and operation. The world's first large-scale tidal power plant, operational in 1966, was the Rance Tidal Power Station in France until Sihwa Lake Tidal Power Station which opened in South Korea in August 2011 which consisted of 10 turbines with a generating capacity of 254 MW. Ocean waves are considered as an indirect form of solar energy and are capable of turning the turbines thus electricity generation as shown in Figure 12.6. The only commercial and operational wave power station, till 2004, is in Scottish island, on its western coast Islay.

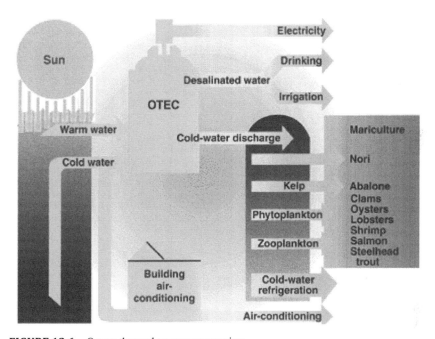

FIGURE 12.6   Ocean thermal energy conversion.

Source: Wikimedia Commons. (https://commons.wikimedia.org/wiki/File:Otec_produk-ty-2_(English).png)

12.4   OCEAN THERMAL

The thermal gradient of the ocean water is used to generate electricity in ocean thermal energy conversion (OTEC) system. This system uses the natural heat of the warm tropical surface waters, where at least a 22°C thermal gradient exists between the surface and the water at the depth of 1000 km, to produce heat which in turn drives the turbines. In order to keep the cycle running, deeper, colder water is pumped to the surface to get heated naturally. This type of ocean energy has many positive and a few negative environmental impacts as shown in Figure 12.7.

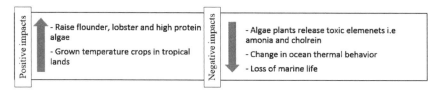

FIGURE 12.7   Positive and negative impacts of OTEC.

12.4.1   IMPACTS OF OCEAN THERMAL

12.4.1.1   POSITIVE IMPACTS

Positive impacts include using the cold pumped water in the mariculture ponds to raise flounder, lobsters, and high-protein algae. This cold water is also successfully being used to grow temperate crops in tropical lands, for example, strawberry cultivation in Hawaii. Industrial cooling and air conditioning of big buildings can be attained using this cold water.

12.4.1.2   NEGATIVE IMPACTS

When implemented on large scale, OTEC plants do leave negative impacts on the environment. Toxic elements like ammonia and chlorine enter the environment from these plants and over the longer time scale, the persistent use of warm surface and cold deep water may lead to a cooling of surface and warming of water at greater depths. This imbalance in ocean thermal behavior may have terrible impacts on the marine life. It can cause the death of corals and fish, leads to lesser hatching rate of eggs, larvae with lesser development. Impingement and entrainment become the major causes of death of marine life.

In addition to marine life, the tropical seas may become hosts of low-nutrient ecosystems, as the cold, high-nutrient water, continues to be used for aquaculture, air conditioning, etc. OTEC plants cause some of the sequestered carbon in the deep cold water, to get outgassed in the atmosphere when the same rise to the surface. This carbon emission is very little when compared with the power generated through fossil fuel burning.

12.5   WAVE ENERGY

Wave energy is one of the richest renewable energy sources which could provide up to 1/5th of current global energy demand (World Energy Council, 1993). Among other renewable energy sources like wind and solar energy which are available for only 20–30% of the time, wave energy is present for almost 90% of the time, at a given site. Since the wave energy is dependent on wind speed, thus the most promising lands with high wave energy potential exist "between 40° and 60° N latitudes, on the eastern boundaries of oceans" (Pelc and Fujita, 2002).

12.5.1   *IMPACTS OF WAVE ENERGY*

12.5.1.1   *POSITIVE IMPACTS*

Wave energy has a large potential in helping the world minimize its reliance on fossil fuels. There are views that by constructing structures of a wave farm, the marine life is provided with structures that act as artificial reefs, which in turn helps to enhance marine life. However, such impact is highly site-selective.

12.5.1.2   *NEGATIVE IMPACTS*

Environmental impact assessment concludes that wave energy can have impacts on the morphology of ocean, its movement, that is, waves and tides, and its fauna and flora. Offshore wave farms also can prove to be lethal for the fish, causing death due to the collision, the changes in the habitat, acoustic trauma, and barrier effects due to electromagnetic effects (EMF) (Pelc and Fujita, 2002). Larger wave energy plants have negative impacts on the marine life and also hamper the ocean atmosphere interactions. Marine mammals become vulnerable to physical entrapment, collision disturbance of habitat, etc., due to the construction of wave power plants. Calmer seas are a requirement of the wave power plants. Achievement of this goal results in slower than the usual mixing of the water in the upper layers of the sea which is harmful for the marine life. The benthic populations are most likely to suffer because of inappropriate mixing of surface water which reduces the food supply to the benthic populations. The larvae of several fish species depend on the currents to be transported from spawning grounds to feeding grounds. As the wave energy devices, alter the currents' routes, thus they negatively affect the fish reproduction. Such plants lead to dampening of the waves, which in turn result in slower shoreline erosion. Wave energy plants also alter the patterns of sediment deposition.

12.6   TIDAL ENERGY

Tidal energy can be regarded as a sustainable future energy source. This is a renewable energy source which is dependent on the sun and the moon's gravitational pull on the waters of the earth, and is also dependent on the centrifugal force which is a result of the earth–moon system. After the successful installation of La Rance tidal barrage, France, in 1967, tidal power plants are being established at other sites with great tidal potential.

The most efficient effort of harnessing ocean energy has been in the form of tidal power. The tidal energy along every coast cannot be used for efficient commercial electricity generation as it requires specific coastal topography, such as found in the north coast of France, north eastern US, etc. In order to harness the energy of the tides a dam is built across the estuary or bay to create a reservoir. One of the best examples is the La Rance tidal power plant in France, which uses the power of the ebb tides to produce about 24,000 KW of electricity from 24 power units. The production of tidal energy is highly dependent on specific coastal and specifically estuary topography. The Bay of Fundy in Canada is the best example of tidal power production, which uses the power of 11-m high tides.

Tides of height above 20 ft. are produced in 30 locations, across the globe, with specific coastal topography. France and Canada have large tidal power plants. Regularity and predictability and immense energy of tides make tidal power more attractive than other ocean energy systems. Harnessing offshore wind energy is also emerging as a powerful renewable energy source. As compared to land, wind speed over the oceans can be 20% higher. Worldwide, total tidal energy potential is about 500–1000 TWh/year (Hammons, 1993).

12.6.1   IMPACTS OF TIDAL ENERGY

12.6.1.1   POSITIVE IMPACTS

The significant environmental processes like nutrient dispersion, the shifting of the pollutants, and sediment transport are all the result of alteration in the transient and residual flow velocities which are caused by the tidal farms. While being favorable for the environment tidal turbines and tidal fences can help in significant power generation. Tidal turbines do not interrupt the migrations of fish and do not tend to obstruct the flow of the channels or block the estuarine mouths.

12.6.1.2 NEGATIVE IMPACTS

The building of a dam is bound to alter the natural ecology of the region, and thus affects the vegetation and wildlife. The catastrophic and rapid change in the habitat of the birds and other marine organisms is due to the periodic filling and emptying of the bay. This is done because the dam is opened and closed with the rise and fall of the tides. Secondly, the dam also has a profound impact on the upstream and downstream passage of the fish.

Building dams along estuaries can cause serious environmental problems, as this is the area where fresh water meets the salt water and become a nutrient-rich nest for many marine species. By the construction of artificial structures, many species of fish and invertebrates are prohibited from reaching their breeding habitats.

Tidal power is attained either through tidal barrages, tidal fences, or tidal turbines. Among these three, most harmful to the marine environment are the tidal barrages, as they have a harmful impact on the water quality and marine life. Since estuaries also serve as great recreational spots, building a dam might finish its scenic beauty and tourist attraction.

Tidal power plants are located at the mouth of estuaries which serve as an ecosystem for many marine species. Their construction results in altering the saltwater flow, a slower rate of current movement, changes in the bottom water characteristics, and reduced salinity. The tidal fences and tidal turbines have lesser harmful impacts on the environment. Tidal fences block the channels, thus obstructing fish migration and the turbines can cause killing of the fish. However, by gearing the turbines at lower velocities, the harm to marine life can be controlled to a certain extent. Tidal farms can harm the fisheries and kill the migratory birds and seabed species. Mammals get disturbed due to the vibrations caused by these plants.

12.7 GEOTHERMAL ENERGY

Raven and Berg, 2004, state that the first 10 km of the earth's crust contains 1% of the heat of the earth, which is equivalent to 500 times the energy that can be produced by all the oil and natural gas reserves of our planet. The United States leads the geothermal power production in the world. Other significant geothermal power utilizers are Japan, Iceland, Italy, Mexico, Philippines, and Indonesia. Geothermal energy is considered environmentally friendly, in sharp contrast to fossil fuel consumption.

Heat within the earth is more toward its center and dissipates near the surface. The geothermal average gradient is 30°C/km of depth. The heat within the earth needs a carrier to reach the surface. This heat reaches the subsurface by the two processes of conduction and convection, through carriers like rainwater. This water reaches hot rocks and forms aquifers with temperatures of above 300°C (Barbier, 2002). Geothermal fields are characterized by such aquifers.

Shallow earth, low-density geothermal energy is not used to produce electricity but is used for heating the buildings, the water in swimming pools, and for maintaining the soil temperature in greenhouses for crop production. Bayer, 2013, states that the total world's installed geothermal power capacity is 13 GW. Among the countries of the world, 90% of the total produced geothermal power is by the United States, followed by Philippines, Indonesia, Mexico, Italy, Iceland, New Zealand, and Japan. According to an estimate geothermal energy has the potential to fulfill 3% of the global electricity and approximately 5% of the global heating demand by the year 2050 (Shortall et al., 2015).

12.7.1 IMPACTS OF GEOTHERMAL ENERGY

12.7.1.1 POSITIVE IMPACTS

The geothermal heat pumps (GHPs) are known for their effectiveness as heating systems and the lowest producers of carbon dioxide, according to the EPA. It releases very minute fraction of air pollutants and disturbs very small amount of land use. Some of the benefits of geothermal energy include that they can lead to easier access to drinking water, through water pumping or drilling of freshwater wells (Fig. 12.8).

12.7.1.2 NEGATIVE IMPACTS

Since the geothermal fields are located in seismically active areas they can cause alterations in local geological conditions. There has been evidence that in high temperature energy fields, due to reinjection and exploitation, the small-*magnitude* earthquakes in the field may rise in frequency, for example, Geysers in California. Subsidence of land surface may result due to the extraction of geothermal fluids from the geothermal energy reservoirs. Depending upon the characteristics of the geothermal field, extraction of geothermal energy results in the emission of the following gases: carbon dioxide, hydrogen sulfide, ammonia, silicates, minerals, volatile matter, carbonates, metal sulfides, and sulfates. Moreover, the local cloud formation and weather is affected to some extent because of the steam releasing from the field. Geothermal power plants can alter the landscape and have an impact on the water quality of the area. The harmful effluents containing chlorides, dissolved chemicals, and metals like boron, aluminum, arsenic, etc., spoil the local water quality. There are examples of excessive salt concentration in geothermal fields, which directly has a negative effect on the environment.

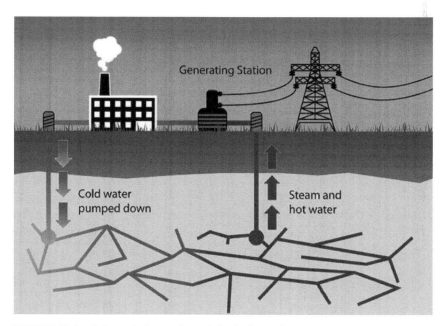

FIGURE 12.8   Schematic for geothermal (hydrothermal) energy.

Source: Rxn111130. (https://commons.wikimedia.org/wiki/File:Geothermal-energy.png)

The rise in the temperature of water, typically leads to decrease in dissolved oxygen, which can be threatening for the marine organisms and lead to unwanted vegetation growth, for example, Waikato river in

Wairakei, New Zealand. Thus, this thermal pollution pushes the organisms to migrate to suitable living conditions and change the biodiversity if the area.

Geothermal energy is notorious for contributing hydrogen sulfide gas, which is the most dangerous environmental hazard this energy source can contribute. Baba and Armannsson, 2006 mention that Turkish geothermal water is loaded with minerals like cadmium, lead, arsenic, and boron. This mineralized geothermal water can cause corrosion and scaling. Geothermal energy also results in the release of nitrogen, hydrogen sulfide, carbon dioxide, ammonia, radon, mercury, and boron. It is not all nature friendly, as it has its harmful impacts for the environment, like the disturbances in the surface caused due to fluid withdrawal, the impact of heat being released, and also the impacts of chemicals. Together these impacts have a negative effect on the biology of the area too.

Noise pollution also results from geothermal energy production; it leaves its impacts on natural habitats and in some cases leads to water pollution. Geothermal plants are a source of noise pollution during their building and drilling and plant operation, etc. Almost up to 120 dB of noise is produced from the drilling of geothermal plants (Bayer, 2013). The groundwater ecosystems are known to be thermally static. Geothermal energy can lead to changes of the temperature in the groundwater ecosystems. This temperature can lead to alteration in the groundwater chemical composition, the metabolism of the organisms, biogeochemical processes, and the functions of an ecosystem. The microorganisms are responsible for natural purification of water. They also help in the biodegradation of the pollutants, the cycling of carbon and nutrient, and the wiping off of the pathogens. The metabolic activities increase with an increase in temperature and vice versa.

12.8   BIOMASS

Energy obtained from the biomass can be referred to as biofuel. Biofuels are classified into three groups, "fire-wood, organic wastes," and crops which are specifically grown to "be converted into liquid fuels." Biomass can be solid, liquid, or gas. Biomass energy sources commonly used are Municipal Solid Waste (MSW), wood and its waste, crops and their wastes, along with their byproducts, food processing waste, animal waste, aquatic plants, and algae. It is burned to generate electricity.

Figure 12.9 depicts the biofuel processing chain. Demirbas et al. 2009 state that almost 35% of the energy demands in the developing countries are met by biofuels. A renewable form of energy, bioelectricity, can be produced from sewage sludge, crops, and MSW. As these sources are rich in organic content thus they ensure great energy potential. Land-fills and incineration are the most commonly used methods to produce bioelectricity. Production of energy by using biomass leads to reduced GHG emissions.

Naik et al. (2010) classify biofuels into two categories first and second generation of biofuels. First generation of biofuels can provide domestic level energy and offer CO_2 benefits. First-generation biofuels are signifi-cant as they can easily blend with petroleum-based fuels, can be used in alternative vehicular technology, can be used in internal combustion engines, etc. While the second-generation biofuels come from plants and are made of a cheap nonfood material that is available from plants. These fuels are carbon neutral and even sometimes carbon negative.

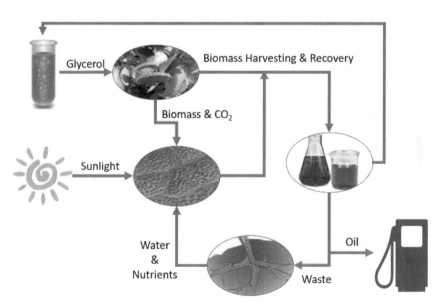

FIGURE 12.9 Biofuel production process.
Source: Adapted from Valente, 2013.

Herbert and Krishnan, 2016 regards biomass as the third largest source of energy in the world, after coal and oil, and state that in 2010, globally

1.5% of world's electricity was produced through biomass. Biomass energy is not entirely environmentally friendly; however, its contribution to GHG emissions is very meager when compared to other fossil fuels.

12.8.1   IMPACTS OF BIOMASS FOR ENERGY

12.8.1.1   POSITIVE IMPACTS

Biomass is a form of cleaner fuel. If this means of energy is sustainably used, it can cause no gain or loss in CO_2 emissions, as the amount of CO_2 released becomes equal to the amount of CO_2 that is consumed during photosynthesis. This means of renewable energy is widely available over the globe even in villages. The main charm of this means of energy is that waste and by-products can be used as fuel. Herbert and Krishnan, 2016, state that biomass provides 10% reduction in the emissions of CO_2. In comparison to the burning of gasoline and coal the combustion of biofuels leads to the production of less harmful and fewer pollutants. Bioenergy is carbon neutral and sometimes can safely be regarded as a carbon sink.

Biomass is capable of using the wastes to produce energy and can reduce the issues related to waste disposal. For example, cow manures can be burnt in furnaces to generate electricity for homes, for example, Mesquite Lake Resource Recovery Project in South California. The areas not suitable for the growth of food crops can be used to grow certain trees with specific nuts, as those are believed to provide energy benefit in such environments. It can help in restoring lands that have been regarded as unproductive, by increasing soil fertility, biodiversity, and enhancing water retention.

There are benefits of using bioenergy as an alternative fuel. It helps developing countries fighting the energy dilemma; it provides energy as liquid and gaseous fuels, and in the form of heat and electricity. Another use of biomass energy is that it is being used on an experimental basis in Australia and Germany as combined heat and power (CHP), which is a new technology which generate electricity and produces useful heat from the same power plant (Renewable 2005 Global Status Report). Ethanol, produced as a by-product of sugars, starch, lignocellulose, and ethanol/gasoline blends are being used as an alternative transportation fuel. Germany and France are good examples in this regard. Biodiesel is also being successfully used as a transportation fuel, thus removing the reliance on conventional transportation fuels. As motor vehicles globally are responsible for emitting 70% of world's carbon and 19% of world's carbon dioxide, thus the use of biofuels as alternative transportation fuel help in reducing GHG emissions (Goldemberg et al., 2008).

12.8.1.2 NEGATIVE IMPACTS

MSW landfills are a great threat to the environment as huge amounts of CO_2 is released from methane released due to its incineration. Similarly, sludge landfills have highly negative impact on the environment as they affect the "terrestrial acidification, marine eutrophication, particulate matter formation… and raw material transportation" and fossil depletion. Heavy metals are released during MSW incineration which leads to harmful effects on "human toxicity and marine ecotoxicity" (Xu et al., 2016). As in fossil fuel combustion, the release of carbon dioxide by biomass burning cannot be negated; however, the quantity is less as compared to the amount of CO_2 generated by fossil fuel combustion. Similarly, biomass combustion generates a lesser amount of sulfur and ash as compared to fossil fuel combustion. Biofuel agriculture can indirectly lead to an addition in the atmosphere of CO_2, as when the natural vegetation is cleared to grow biofuel crops. The use of biofuels can result in land degradation and poor air quality.

Biomass energy leads to air pollution which contributes to global warming. The rate of aerosol release is high during biomass combustion period, which leads to poor visibility. The burning of biomass for heating or cooling purposes produces NO_x, carbonaceous smoke, and particulate matter.

In developing countries there is a heavy emphasis on growing the energy crops, which in turn leads to fluctuation in the food prices. The authors state that globally 2.4 billion people rely on biomass for cooking. If energy crops are to be grown on the agricultural land, this would mean lesser space for food crops and a rise in their prices (Raven and Berg, 2004).

At least half of the world's population relies on biomass, and burns firewood faster than it is being replanted. This exploitation of forest wood leads to a chain of environmental issues like, soil erosion disruption in water supplies, etc. Crop residues like stalks from corn, wheat, etc., are used for energy generation, however, if left on the ground, they can help in soil protection and soil nutrient regain. By constant removal of crop residues, the soil can be deprived of its nutrient replenishment. Biomass removal from fields leads to loss of topsoil by erosion, a decrease in soil fertility, and an increase in runoff. There exists a "low energy to weight ratio of biomass," which leads to its less usefulness as an energy resource. Similarly, the forests can lose their precious soil, due to large-scale harvesting of biomass from forests (Smith, 2008).

Producing energy using biomass can also lead to increased soil erosion. In order to produce energy crops, like ethanol, marginal lands are mostly cleared and used for relevant crop cultivation. This marginal land is sensitive to soil erosion. Also this soil erosion intensifies water runoff, thus minimizing the groundwater recharge. This nutrient-rich runoff ends up in the lakes, or estuaries, causing eutrophication there.

Biofuel agriculture requires an intensive use of pesticides and fertilizers which when mixed with rainwater can prove to create a havoc in aquatic environments. Monocrop agriculture focused on producing biofuels leads to less variance in vegetation, thus contributing significantly to habitat loss. The reliance on biomass in some of the developing countries leads to desertification which might lead to loss of biodiversity. Biomass crop production on a large scale can threat natural habitat of various species as the natural vegetation has to be removed in order to plant biomass crops.

When solid waste is burnt at the household level, without proper ventilation, it leads to severe health issues. The children are known to suffer from respiratory diseases due to the addition of particulates from the burning of biomass. Carbon monoxide is associated with problems during pregnancy. High exposure rates to biomass and coal burning can also lead to cancer. Sulfur, trace metals, and nitrogen oxides all have dangerous impacts on human health. While producing agriculture and forest biomass, the rate of injuries and illnesses of associated professionals are many times higher than those associated with mining coal and oil.

12.9   WIND ENERGY

Wind energy is considered as a clean energy source. Wind turbines are used to transform wind energy into electricity. Wind turbine system mainly depends on wind velocity, wind directions, turbulence intensities, and roughness of the surrounding terrain. Its operational lifetime is 20–25 years (Klugmann-Radziemska, 2014).

From ancient times, wind energy has been used as one of the primary energy sources for transporting goods via Nile river, milling grain in Persia, and pumping water in China and generation of electricity in the early 20th century in Europe and North America, etc. Some of the well-known regions of wind power development areas are shown in Figure 12.10.

In the history of wind energy evolution, it was steadily declined also because oil and coal have been proven to be highly effective drivers of economic growth in Europe and America at that time; however, the oil crisis of the 1970s promoted rebirth of the wind industry. Later in the 1990s it accelerated with rapid pace in the United States and Europe, after the installation of the first modern grid-connected wind turbines. At that time, it generated 2000 MW which has not expanded to 486,790 MW. In the past, the growth of wind energy is seen more particularly in Europe as these countries were more concerned about sustainable electricity generation. But since 2006, Asia has picked up steam and now possesses installed capacity (MW) which exceeds Europe. Trend of wind energy restarted in Asia and now led by China who is producing electricity by wind power over 80% in Asia (GWEC, 2006; Statistics, 2015). Thus, global installed capacity of wind power is highest in Asia now as shown in Figure 12.11.

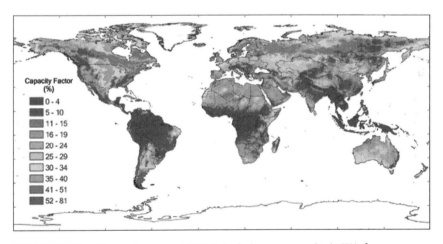

FIGURE 12.10   (See color insert.) Global wind power capacity in W/m^2.

Source: National Academy of Sciences of the United States of America. (https://www.pnas.org/content/106/27/10933)

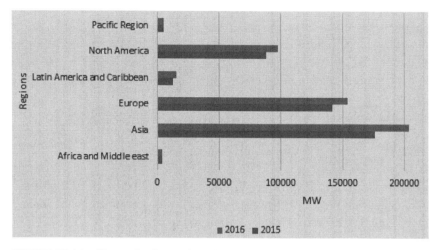

FIGURE 12.11   (See color insert.) Global installed capacity (MW) of wind regional distribution.

Source: Reprinted with permission from REN21, 2017.

12.9.1   *IMPACTS OF WIND ENERGY FARMS*

12.9.1.1   *POSITIVE IMPACTS*

Wind energy is infinite and well-known environment friendly resource because of its minute contribution in GHGs emission. It behaves as a viable resource and can be harvested either in the mainland or on the ocean. More-over, its installation and transmission are possible and accessible even in remote areas with never-ending supply. Moreover, the electricity produced by these systems could save several billion barrels of oil and avoid many million tons of carbon and other emissions. Similarly, it secures water as well. For instance, conventional means of power generation require lots of water. The amount of water utilized by thermoelectric power plant range from 20,000 gallons per megawatt hour to 60,000 gallons per megawatt hour for electricity generation. However, wind energy is a great way to conserve freshwater and wind turbines do not need water once started to operate (Logan et al., 2012). Wind's environment friendly nature also contributes to a great reduction in CO_2 emission during its entire process of electricity generation.

12.9.1.2 NEGATIVE IMPACTS

Although wind is a clean energy source and has many environmental benefits, at the same time its environmental costs cannot be ignored. Acoustic noise emission, visual impact on the landscape, impact on bird's life, and shadow flickering are the main environmental disturbances caused by the rotor, and electromagnetic interference influencing the reception of radio, TV, and radar signals.

The first drawback of wind turbines is noise pollution. Noise generated by a wind turbine can be divided into two broad categories: (1) mechanical and (2) aerodynamic. Mechanical noise is produced by the moving components such as gear box, electrical generator, and bearings. Normal wear and tear, poor component designs, or lack of preventative maintenance may all be the factors to create mechanical noise. On the other hand, aerodynamic type of noise is developed by the flow of air over and past the blades of a turbine. Such a noise tends to increase with the speed of the rotor and can be heard at distances at least as great as 2.5 km under some conditions (De Vries et al., 2007).

It is evident that level of noise pollution fluctuates by wind turbine under different meteorological conditions. For instance, the sound was louder than average when the strong wind was blowing down from the wind turbines and toward the dwelling, particularly during night time. This creates annoyance in surrounding dwellings. The noise limit for wind farms is 35 A-weighted decibels. Therefore, the proportion of respondents who were annoyed by the sound level up to 35 to more than 45 dB (McCunney et al., 2014).

However, noise emissions of both mechanical and aerodynamic turbine types are now controlled by aeroacoustic designs which divert wind which significantly reduced the level of infrasound of modern wind turbines.

Some direct, that is, mortality of birds by collision and indirect impacts, that is, habitat disruption and displacement have also been reported by different researchers in recent years. Mortalities or lethal injury of birds can result not only from collisions with rotors but also from collision with other structural components of the wind turbine. Bats, golden eagles, swans, red-tailed hawks, American kestrels, Cantabrian capercaillie, and raptors are notable species affected by wind turbines all over the world. It has also been estimated that 234,000 birds on an average were killed annually by collisions with monopole wind turbines in the United States (Dai et al., 2015). Emitted light from wind turbines, impact low-flying avian during poor weather (such as fog, rain, strong wind, or dark nights) and may face mortalities. Among all the types of turbines, bird mortalities are recorded highest due to lattice turbines. It has also been reported that noise produced by the wind turbines may scare birds away. For instance, prairie birds were distracted by at least 100 m by avoiding power lines and road construction sites (Pruett et al., 2009). However, in comparison of adverse impacts of fossil fuels on wildlife, the ratio of bird mortalities resulting from wind turbines are insignificant. These minor impacts can also be further reduced by improvements like the use of avian radar technologies (which can detect birds as far as four miles from turbines), advanced tower design, and minimal use of light (for making the turbines less attractive for migrating birds).

Lastly, visual impacts during days with a clear sky, wind turbines can be seen from as far as 30 km and delineate the zone of visual influences (ZVIs) (Bishop, 2002). Problems associated with visual impacts vary due to wind energy system and technology itself such as wind turbine color or contrast, size and distance from the dwellings in surroundings. Among visual impacts, shadow flickering is observed first and foremost which can be annoying when the shadow of moving turbine blades fall on a house at certain times of the day and year. Some people may look at the wind turbines as machines that are damaging the natural beauty of landscape into an industrial environment. However, this problem can be overcome through Geographic Information System (GIS) or turning the turbine off for a few minutes of the day when the sun is at the angle that causes flicker.

12.10   SOLAR ENERGY

Energy from the sun is harnessed using photovoltaic (PV) cells and thermal technologies with particularly focused on the PV system to get unlimited and clear supply. Solar energy is converted to electrical power from solid-state devices in solar PV technology, while solar thermal generates or concentrates heat for direct distribution into pipes or to generate steam at large scale.

PV cells are usually linked in series or parallel combinations within modules to form arrays for off-grid or grid-connected systems for power generation. Solar thermal or concentrating solar power (CSP) systems are also used for electricity generation in addition to PV systems. In CSP systems, concentrated solar radiation is converted to thermal energy first before it is converted to electrical energy. A CSP plant comprises solar collectors, receivers, and a power block. The solar collectors concentrate sunlight to raise the temperature of transfer fluid in the receiver. The heated fluid produces steam to generate electricity through a conventional turbine. Many studies show that PV power systems will have an important share in the electricity of the future. While it seems that solar PV has a promising future, a critical analysis of the solar energy technologies is also skeptical and assumes that these should be seen as supplementary resources and not as alternatives. The Australian continent has the highest solar radiation per square meter of any continent and consequently some of the best solar energy resources in the world. The regions with the highest solar radiation are the desert regions in the northwest and center of the continent (Australia, 2010).

12.10.1 IMPACTS OF SOLAR POWER PLANTS

12.10.1.1 POSITIVE IMPACTS

Solar energy technologies have certain environmental benefits compared to energy generation from traditional fossil fuels. The main advantage is a reduction in CO_2 emissions, the absence of any air emissions or waste products during operation. Other advantages include reclamation of degraded land, reduction of the required transmission lines of the electricity grids by distributed generation and improvement of the quality of water resources. Solar systems have the potential to reduce environmental pollution. The environmental protection is offered by the widely used solar water heating and solar space heating. The use of solar energy provides considerable savings in greenhouse polluting gasses. In the case of domestic water heating systems run with electricity or diesel backup, the saving, compared to a conventional system, is about 80% (Kalogirou, 2004).

Solar and other renewable energy technologies can provide small incremental capacity additions to the existing energy systems that usually provide more flexibility in incremental supply than large units such as nuclear power stations. Sometimes, renewable energy technologies are seen as direct substitutes for existing technologies so that their benefits and costs are conceived in terms of assessment methods developed for the existing technologies. Development of renewable energy technologies can serve as cost-effective and environmentally responsible alternatives to conventional energy generation.

Apart from that, solar water heating has the benefit that it heats water without producing harmful emissions into environment compared to other energy-intensive fossil fuel-based boilers. Thus, near zero carbon dioxide emissions appears to be the greatest environmental impact that falls to the benefit of the solar energy technologies. As for the major socioeconomic benefits of solar energy technologies, an increase of the national energy independence, diversification and security of energy supply, and rural electrification in developing countries have a positive impact.

12.10.1.2 NEGATIVE IMPACTS

While solar energy is a renewable resource and it is expected to counter the adverse effects of climate change, there are some negative aspects to the environment that have brought this clean technology into critical scrutiny. It is generally accepted that solar energy is diffused and not fully accessible, intermittent due to cloud cover and limited sunshine, distinct regional variability due to shadowing and latitude effect.

The other undesirable effects of solar energy technologies are usually minor and can be minimized by appropriate mitigation measures. The potential environmental impact is usually site-specific, depending on the size and nature of the project. The negative impacts may include, but are not limited to, the noise and visual impact, GHG emissions, water and soil pollution, energy consumption, and labor accidents during the construction and installation phase. The study also notes the environmental impact on important archaeological sites or on sensitive ecosystems especially in the case of the central solar technologies. Among the environmental impacts of solar thermal technologies, this study also lists the accidental leakage of coolant systems leading to fire and gas releases from the vaporized coolant, adversely affecting the public health and safety. There is also visual impact on buildings' aesthetics as a negative impact, however, adoption of standards and regulations for environmental friendly design along with good installation practices and improved integration of solar systems in buildings could lead to little impact.

Solar technology is much preferable to traditional means of power generation while also considering wildlife and land use impacts. All high-priority impacts are favorable to solar power displacing traditional power generation, and all detrimental impacts from solar power are of low priority. The impact on plant and animal life is a major hurdle for permitting the construction of the large solar power plants. Also, impact on water resources for cooling of the steam plant and, possibly, water pollution due to thermal discharges or accidental discharges of chemicals used by the system may occur but appropriate constraints and improved technology can eliminate risks.

Although solar energy experts select feasible sites, they frequently encounter conflict over biodiversity conservation values that were not factored into the initial suitability rating methods. Some of the environmental impacts can be mitigated, such as those for biodiversity which includes clearing land to install ground-mounted PV systems.

Yet another negative impact that has little likelihood but should be mentioned is the risk of chemical waste. During their normal operation, PV systems do not emit any pollutants or radioactive substances. However, in CIGS (copper indium gallium selenide) and CdTe (cadmium telluride) modules, which include small quantities of toxic substances, there is a potential slight risk that a fire in an array might cause small amounts of these chemicals to be released into the environment.

12.11 CONCLUSION

Energy continues to be the major need for the development of any country. The energy crisis is the one that made mankind strive for more sustainable, that is, renewable forms of energy. This chapter shared some of the impacts of the renewable energy forms on the environment. Positive impacts though being lesser as compared to the negative ones, when compared with the conventional sources of energy dependent on fossil fuels, all the sources of renewable energy, that is, solar, wind, hydropower, ocean, geothermal, and biofuels proved to be lesser harmful to the environment. In most cases, the renewables proved to be carbon neutral and in some cases carbon negative even. It is thus concluded that the renewable sources of energy are not all environment friendly, as they do have a significant impact on biodiversity, soil, aquatic environments, marine life, air quality, and even human health. Further research is suggested to be carried out regarding the negative impacts of renewable sources of energy so that their efficiency and reliability may be further affirmed for future.

KEYWORDS

- renewable energy resources
- geothermal
- solar
- wind
- hydropower
- ocean energy
- biomass

REFERENCES

ABARE. *Australian Energy Resource Assessment*, Geoscience Australia, Canberra, 2010.

Bishop, I. D. Determination of Thresholds of Visual Impact: the Case of Wind Turbines. *Environ. Plann. B Plann. Des.* **2002,** *29* (5), 707–718.

Dai, K.; Bergot, A.; Liang, C.; Xiang, W.-N.; Huang, Z. Environmental Issues Associated with Wind Energy: A Review. *Renew. Energy* **2015,** *75,* 911–921.

De Vries, S.; Lankhorst, J. R.-K.; Buijs, A. E. Mapping the Attractiveness of the Dutch Countryside: a GIS-based Landscape Appreciation Model. *Forest Snow Landsc. Res.* **2007,** *81* (1), e2.

Gielen, D. Renewable Energy Technologies: Cost Analysis Series. *Sol. Photovolt.* **2012,** *1* (1), 52.

Giesecke, J.; Mosonyi, E. *Water Power Plants*, 4th ed.; Wasserkraftanlagen: German, Springer: Berlin, 2005.

GWEC. *Global Wind 2005 Report*, 2005. Available at http://www.gwec.net (accessed Nov 17, 2017).

GWEC. *Global Wind 2006 Report*, 2006. Available at http://www.gwec.net (accessed Mar 6, 2018).

Kalogirou, S. A. Environmental Benefits of Domestic Solar Energy Systems. *Energy Convers. Manage.* **2004,** *45* (18), 3075–3092.

Klugmann-Radziemska, E. *Environmental Impacts of Renewable Energy Technologies*. International Proceedings of Chemical, Biological and Environmental Engineering (IPCBEE), 2014, 69, 104–108.

Logan, J.; Heath, G.; Macknick, J.; Paranhos, E.; Boyd, W.; Carlson, K. *Natural Gas and the Transformation of the US Energy Sector: Electricity*. Joint Institute for Strategic Energy Analysis, 2012.

McCunney, R. J.; Mundt, K. A.; Colby, W. D.; Dobie, R.; Kaliski, K.; Blais, M. Wind Turbines and Health: A Critical Review of the Scientific Literature. *J. Occupat. Environ. Med.* **2014,** *56* (11), 108–130.

Pruett, C. L.; Patten, M. A.; Wolfe, D. H. Avoidance Behavior by Prairie Grouse: Implications for Development of Wind Energy. *Conserv. Biol.* **2009,** *23* (5), 1253–1259.

Statistics, G. W. (2016). Global wind energy council. Washington, DC, USA. URL: https://www.gwec.net/wp-content/uploads/vip/GWEC_PRstats2016_EN_WEB.pdf

Twidell, J.; Weir, T. *Renewable Energy Resources*, 3rd ed. Routledge: London, 2015.

Valente, V. (2013). InteSusAl, sustainable microalgae biofuel production visible to the naked eye. http://www.besustainablemagazine.com/cms2/intesusal-sustainable-microalgae-biofuel-production-visible-to-the-naked-eye/

von Sperling, E. Hydropower in Brazil: Overview of Positive and Negative Environmental Aspects. *Energy Procedia* **2012,** *18,* 110–118.

CHAPTER 13

Geological Influence on Surface Water Quality

LOO MEI YEE and SANDEEP NARAYAN KUNDU*

*Department of Civil & Environmental Engineering,
National University of Singapore, Singapore*

Corresponding author. E-mail: snkundu@gmail.com

ABSTRACT

Water occurs as surface water and groundwater. Contamination of water
from various sources impact the quality of water and obliterates it for
general use. Apart from anthropogenic contamination, water bodies are
naturally contaminated too from the minerals in the rocks they are hosted.
Water, which is otherwise neutral, is mostly found to carry some acidity
or basicity resulting from the contamination. The following chapter evalu-
ates the geological influences on water bodies in similar settings but with
different underlying geological materials. It discusses the testing of water
to detect such influences and determines the quality of water for specific
use in urban society.

13.1 WATER AND ENVIRONMENT

Water is a critical natural resource which is required by plants and animals
alike to survive and flourish. Water has many uses apart from drinking,
which includes the likes of domestic use (cooking, bathing, washing,
in swimming pools, and watering plants in gardens) and industrial use
as in agriculture, power production, and other industrial uses. Potable
water is the water suitable for drinking and is transparent and free from
contaminants like germs and chemicals. Contaminants are introduced into
water bodies naturally and through anthropogenic activities and alter its
quality, rendering it unfit for human and industrial use. There have been

several occasions where epidemics have resulted from the consumption of nonpotable contaminated water. Water in different environments has different contaminants. Routine checking of water resources is necessary to identify the contaminants and effect necessary treatment for rending it usable for humans and society.

Water is generally classified as surface water and groundwater. The amount of precipitation which percolates into the ground and occupied the pores of rocks is called the groundwater. The geological formations which are porous and permeable and contain water are called aquifers. Groundwater is seldom in direct contact with the atmosphere. Surface water is generated from runoff resulting from precipitation. Rainwater flows toward basins in a confined area called the catchment. In a catchment, rainwater feeds holding areas like lakes and reservoirs via rivers and streams. Catchments for surface water bodies are large areas which are mostly far away from the urban regions which minimize the chance of the water being polluted. Both surface water and groundwater are important resources which are critical to sustaining ecosystems on earth. Surface water is readily available for human use, whereas groundwater needs to be extracted using a well. Traditionally, most of the surface water has been used for irrigation and with increased urbanization, thermal power generation utilities have been using this resource for cooling machinery. By 2010, in the United States, 50.4% of the total surface water withdrawals were used this way with only 29% accounted for agriculture.

The environment surrounding the surface waters bodies impact the quality of water. Various contaminants from organic to inorganic origins are found in water bodies. Surface water bodies are not usually very high in mineral content which is why it is called soft or freshwater. Its exposure to the atmosphere above and the rocks below triggers interactive processes and chemical exchanges which results in many contaminants in suspended and dissolved form in the water. Contaminants are present even in pristine mountain streams which are sourced from wild animal and plants and weathering of surrounding rocks. Despite being cut off from the atmosphere, groundwater has contaminants too. This is because it is recharged from surface water bodies which bring in contaminants into the groundwater but the majority of it is received from the weathering of the host rock itself. However, groundwater contains fewer contaminants than surface water as the rock acts as a filter removing several contaminants. In this chapter, we shall look at the various contaminants in water bodies and focus on the influence of bed-rock geology on surface water bodies, especially lakes.

13.2 CONTAMINATION OF WATER

Water bodies are in contact with soil, rocks, and the atmosphere. Therefore, geology and soil properties affect water quality. Water contaminants come from three primary sources, namely, rainwater, surface water, and groundwater. Rainwater is in general safe to drink, but has become increasingly contaminated because of pollutants in the atmosphere which are picked up by precipitation. Anthropogenic land use is one prime reason for increased air pollution in many parts of the world. Surface water bodies are fed by precipitation runoff which picks up contaminants from the catchment area. Land cover and land use of the catchment area based on agricultural or industrial practices contribute to the pollution in rivers, streams, and lakes. Water from underground aquifers is usually safe to drink, but increased pollution in groundwater recharge areas have resulted in increased contamination. Groundwater in deeper aquifers are more protected from many types of contamination but chemical contamination in the form of elements like arsenic, selenium, or boron from the hosted aquifer is common. Water contaminants fall into four basic categories. These are pathogens or germs, suspended solids, dissolved salts, and chemicals.

13.2.1 BIOLOGICAL

Pathogens in water pose a serious health risk for everyone in the community. Pathogens include the likes of bacteria, cysts, parasites, and viruses which when introduced into the human body through potable water can cause diseases and even epidemics. All potable water sources such as bores, rivers, lakes are regular for these pathogens and necessary treatment is done before the water can be declared safe for drinking. All water sources except rainwater collected with germ-free equipment are susceptible to biological contamination and the chances of such contamination is high in tropical climates.

13.2.2 SUSPENDED SOLIDS

Suspended solids include particulates present in water which preserved their physical state and reside in the suspended form. These include clay minerals, iron oxides, or organic matter. Suspended solids impart an offensive taste and odor to the water. These solids are mostly sources from natural sediments and unnatural sources like industrial effluents which are carried by the runoff to the water bodies. Suspended solids impact the aesthetics of water and can be removed by decantation, a process in which the suspended matter is allowed to settle and then filtered through a screen or membrane.

13.2.3   DISSOLVED SALTS

Water being a good solvent takes much material into solution as it passes from the catchment to the reservoir. The underlying geology influences the kind of salts found in water bodies. Various minerals like salts of sodium, calcium, and magnesium are chemically weathered by running water and end up in solution in water bodies. Oceans are hosts to all the runoff from land and contain huge amounts of salts, which is why ocean waters are saline. The common salts found in water bodies are sodium chloride (common salt), calcium carbonate (limestone), and magnesium sulfate. The concentration of the dissolved salts makes the water hard which when boiled gets deposited as the white crust in kettles and toilet cisterns. Hard water is neither potable and has a limited industrial use. Should the concentration of salts be above the permissible limit, it needs to be treated chemically. However, this is not often done. Heavy metals like iron, manganese, lead, mercury, arsenic, and cadmium are also present in this water, small concentrations of which can pose a serious health hazard.

13.3   WATER QUALITY

Water quality refers to the chemical, physical, biological, and radiological characteristics of water (Diersing, 2009). It is a measure of the condition of water relative to the requirements of one or more biotic species and/or to any human need or purpose (Johnson, 1997). Water quality is measured against the standards which may be different from country to country with the most common ones being used to assess water quality against its usage types like drinking, washing, and industrial manufacturing.

13.3.1   QUALITY STANDARDS

Water quality standards are the foundation of the water quality definitions which is defined by most countries through national water acts. Quality standards form the legal basis for governance and controls on the amount of contamination of water sources such as domestic supply, industrial facilities, wastewater treatment plants, and sewerage networks. In summary, water quality standards are developed to monitor and maintain water quality necessary for its various usages in a society which range from swimming, recreation, public water supply, and/or aquatic life to human communities. These standards are referred to while sourcing water for a particular use from the surface or ground water reservoirs and should the quality be below par, appropriate treatment options may be weighed upon before water usage from such sources.

13.4 GEOLOGICAL INFLUENCES

Rock-water interaction is an important component of natural geochemical cycling. Minerals in rocks contribute to matter in suspended and solution in water bodies. These constitute the material exchanges from solid phase of rock minerals to the liquid phase in the water system. The dissolved ions and elements are supplied by the rocks in the catchment which in most cases is also the bedrock of the water body. Geochemistry and mineralogy of the bedrock is therefore an important contributor to the water chemistry and quality.

Water is an important agent in weathering and influences the weathering and alteration in the bedrock. In tropical climates, the rate of weathering is accelerated due to high precipitation and temperature which is why water bodies tend to be more contaminated in the tropics than that temperate and polar regions. Contaminants in solution are a result of chemical weathering and the type of elements exchanged depends on the mineralogy of the bedrocks. Bedrock geology plays a very important role in natural contamination of surface water bodies.

13.4.1 REGIONAL SETTINGS

Apart from natural surface water bodies, many are artificial ones like lakes resulting from impounding of rivers by dam projects. Surface water bodies are a source for domestic and industrial water supplies and hence their regional setting is important for defining the usage of its waters. Being a part of the hydrological framework of the region most surface water bodies are connected to rivers. Depending on the stage of the river, the river bed may consist of loose sediments of hard rock. Stagnant water bodies like ponds and lakes have different physiochemical characteristics than flowing water bodies like rivers. Important physical properties of the water are temperature and hydrogen-ion concentration (pH) which mostly depends on the regional settings of the water body. pH of water is an index of the environmental conditions of the region. The water tends to be more alkaline if it possesses more carbonates and less alkaline when it possesses large quantities of calcium, bicarbonates, and CO_2 (Sarwar and Wazir, 1991).

13.4.2   BASEMENT ROCKS AND TYPES

The contaminants of water differ if the bedrock is granite or basaltic as the mineralogy of the two rocks are very different. Basalt is an alkaline rock whereas granite is acidic. The constituent minerals have differential susceptibility to chemical weathering and therefore, result in varying concentration of different cations and anions in water bodies. The bedrock of most stagnant water bodies is mostly nonsedimentary and therefore, the constituent minerals influence the water quality primarily because of the chemical weathering. Most hard rocks are igneous and are classified on the basis of mineralogy, chemistry, and texture. The texture is dependent on whether the igneous rock is the plutonic rocks (large-grained minerals) or volcanic (glassy minerals). In porphyritic rocks, which contain both plutonic and volcanic rocks, the mineralogy greatly varies with subsurface depth. Volcanic rocks are less likely to host water bodies except for hydrologically closed crater lakes.

QAPF diagrams are used to classify plutonic rocks for which mineralogy can be easily identified from hand specimens. A QAPF diagram (Fig. 13.1) is a diamond-shaped diagram, vertices of which represent four mineral end members. Each rock is analyzed for the normalized percentages of these minerals and plotted in the diagram to arrive at the nomenclature. The acronym QAPF stands for the four minerals; Q for quartz (a mineral resistive to all forms of weathering), A for alkali feldspar, P for plagioclase, and F for feldspathoids or foids.

13.4.3   ROCK WEATHERABILITY

Weathering is the disintegration of soil and rock materials at and near the earth's surface by physical, chemical, and biological processes (Ceryan et al., 2007). The main factors that govern weathering of rocks are the climatic and the geographic conditions, the composition of rocks, the groundwater condition, and the period of time exposed to weathering (Zhao et al., 1994). Physical weathering generates particulates which either settle at the bottom of surface water bodies or stay suspended in the water itself, adding to the turbidity. The hardness of the minerals in the bedrock and the distance of transportation define the size and shape of the particulate sediments. Rainwater is slightly acidic and several minerals being susceptible to forming solutions aid chemical weathering.

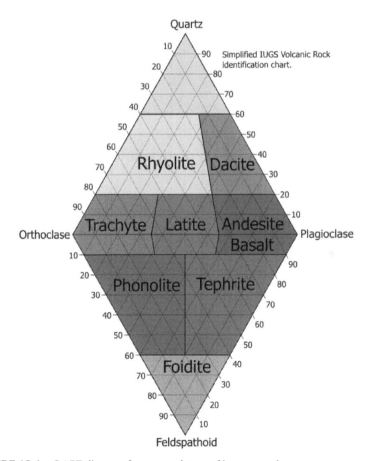

FIGURE 13.1 QAPF diagram for nomenclature of igneous rocks.

Source: USGS. (http://www.geologyin.com/2015/08/using-qapf-diagram-to-classify-igneous.html)

Goldich (1938) presented an order stability of rock-forming minerals which is, in essence, the reverse of the Bowen's reaction series. The order for the rates in which the chemical constituents are altered and eroded by weathering processes is on the basis of stabilities of minerals in different temperature and pressure regimes. Mafic minerals, which are rich in magnesium silicates, are less stable than the felsic minerals (potassium aluminum silicates). Common minerals in igneous rocks crystallize from the melt in two series in order of increasing stability (Fig. 13.2). Olivine has the lowest stability and its rate of weathering exceeds other common ferromagnesian silicates. Within the mafic group, the stability increases from olivine onwards followed by augite, hornblende, and biotite (most stable).

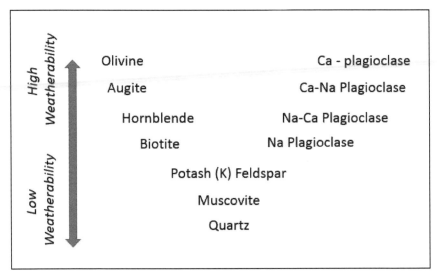

FIGURE 13.2   Mineral stability series in weathering.

A comparative study of gabbro and granite weathering (Fritz, 1988) compared major minerals common to both lithologies, especially plagioclase and biotite. The results showed little relative depletion of major elements in gabbro weathering as compared to granite, contradicting to the Goldich's stability. This is suggestive of different weathering patterns of similar minerals from different bedrock lithology. Ollier (1984) on the other hand proposed that weathering series can be related to basicity, which is the ratio of silica and the other cations. Increased concentration of cations trigger bonding with hydrogen in the water thereby increasing the weatherability of the rock.

These reactions contribute to the ions in surface waters which also react with available carbon and oxygen from the atmospheric interactions to form sediments which may precipitate to the bottom of the water bodies.

13.5 ANALYSIS OF WATER

Of all the water bodies, oceans are spread over 71% of the earth's surface and hold 97% of earth's total water. Being saline, this water is normally not usable for human use unless desalinated. The remaining 3% water is available for humans and occurs in the form of surface water bodies on land and groundwater. On land, surface water and ground water therefore form the critical resource which caters to the use of humans and animals alike. Contamination of this important resource can render it unusable which can disrupt normal life on earth. It is therefore important that the constituents of these waters are analyzed and monitored over time to ensure that its quality is sustained and therefore available for our use. Water is analyzed for its quality and is tested for its instantaneous physical properties, anions, and cations which can render the water hard or soft and organic matter (carbon).

13.5.1 PHYSICAL PROPERTIES

Physical properties are parameters which are normally measured onsite and include the likes of temperature, pH, specific conductance, conductivity, total dissolved solids (TDS), and dissolved oxygen. These are instantaneous in situ measurements which can be performed by commercial research instruments. Temperature is a climatic condition but pH depends on the ionic content in the water which is accumulated by exchanges between the water and the surrounding environment and ecology. Conductivity and specific conductance reflect on the solutes in the water and TDS indicate the turbidity of the water. Dissolved oxygen indicates the health of the water and its supportability of life which is also interdependent on the solutes and turbidity of the water and may vary with depth in the water body.

An examination of two adjacent rock quarry-hosted lakes in Singapore showed significant differences in their pH which was attributed to silicate weathering of the host rocks. One lake was hosted predominantly by gabbro whereas the other had a granite base. Silicate weathering is the most important pH-buffering mechanism in the absence of carbonate minerals. As shown in Table 13.1, the silicate weathering reactions use up hydrogen ions which adds to the pH while releasing cations and silica into the water (Appelo and Postma, 2005).

TABLE 13.1   Equations of Weathering Reactions of Some Silicate Minerals (Appelo and Postma, 2005).

Silicate minerals	Kaolinite
Sodic plagioclase (albite)	
$2Na(AlSi_3)O_8 + 2H^+ + 9H_2O$	$\rightarrow \quad Al_2Si_2O_5(OH)_4 + 2Na^+ + 4H_4SiO_4$
Calcic plagioclase (anorthite)	
$Ca(Al_2Si_2)O_8 + 2H^+ + H_2O$	$\rightarrow \quad Al_2Si_2O_5(OH)_4 + Ca^+$
K-feldspar (microcline)	
$2K(AlSi_3)O_8 + 2H^+ + 9H_2O$	$\rightarrow \quad Al_2Si_2O_5(OH)_4 + 2K^+ + 4H_4SiO_4$
Pyroxene (augite)	
$(Mg_{0.7}CaAl_{0.3})(Al_{0.3}Si_{1.7})O_6 + 3.4H^+ + 1.1H_2O$	$\rightarrow \quad 0.3Al_2Si_2O_5(OH)_4 + Ca^+ + 0.7Mg^{2+}$ $+ 1.1H_4SiO_4$
Mica (biotite)	
$2K(Mg_2Fe)(AlSi_3)O_{10}(OH)_2 + 10H^+ + 0.5O_2$ $+ 7H_2O$	$\rightarrow \quad Al_2Si_2O_5(OH)_4 + 2K^+ + 4Mg^{2+}$ $+ 2Fe(OH)_3 + 4H_4SiO_4$

Conductivity of water is measure of its ability to conduct electricity. Conductivity increases with the amount of dissolved ions in the water. It increases with increase in TDS (Fig. 13.3) and decreases with decrease in temperature (Boyd, 2015).

Gabbro has more unstable metallic minerals compared to granite; its weathering contributes to the increased conductivity of the waters. Gibbs diagrams are plotted to determine the processes controlling the dissolved chemical constituents (Fig. 13.4). The plots indicate whether the dissolved ions in water result from evapotranspiration processes, precipitation, or rock weathering. For the samples collected from the two rock-hosted lakes, rock weathering was found to be the predominant processes with the levels of TDS varying with the host rock weatherability (gabbro being more weatherable than granite).

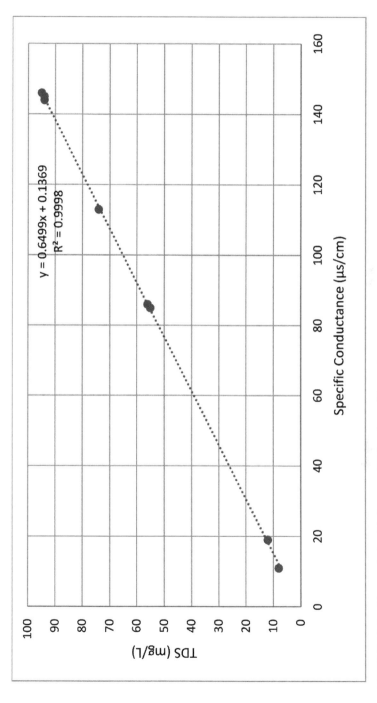

FIGURE 13.3 Linear relationship between TDS and specific conductance.

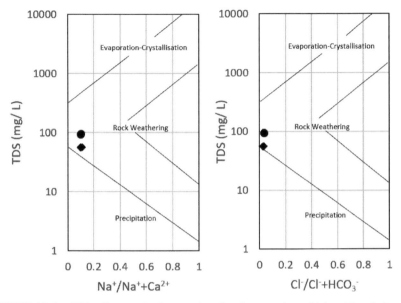

FIGURE 13.4   Gibbs diagram for the samples of rock quarry-hosted lakes. The circle and diamond represent the gabbro and granite-hosted lakes, respectively.

13.5.2   MAJOR IONS AND HARDNESS OF WATER

Presence and absence of major ions in the water are indicative of the water quality. Based on the concentration of the main cations (Na^+, K^+, Ca^{2+}, Mg^{2+}) and anions (Cl^-, SO_4^{2-}, HCO_3^-), a Piper diagram is plotted (Fig. 13.5). The plots can indicate the hardness of water based on different ion types as per the categorization of water type by Piper (1953). Carbon hardness may be present even if the host rocks do not contain any carbonate rocks (as in the case of the Singapore lakes) and this was attributed to weathering of silicate minerals which fixed atmospheric carbon dioxide forming bicarbonates which form weak acid. The presence of weak acid leads to alkalinity and in the case of Singapore as the balance between weak acid and alkaline earths presents secondary alkalinity or temporary hardness (Rogers, 1919). The temporary hardness of water means that the hardness can be removed by the precipitation of calcium and magnesium carbonates (Boyd, 2015). Piper plots are also used for source rock deduction which is helpful in case the provenance of solutes is in the vicinity of the lake and not directly underneath it.

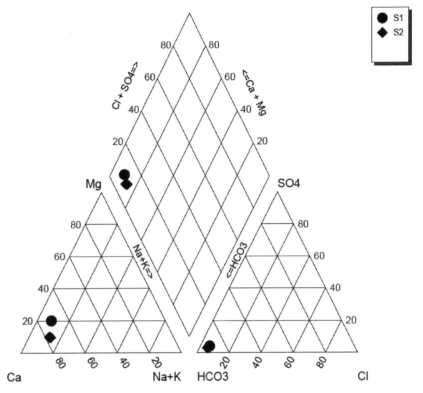

FIGURE 13.5 Piper's plot for the water samples from quarry-hosted lakes of Singapore. The circle indicates the gabbro-hosted lake and the diamond represents the granite-hosted one.

13.5.3 ORGANIC AND INORGANIC CARBON

Another important analysis is that of the total dissolved carbon in the waters. Dissolved carbons can be of organic or inorganic origin and can be differentiated by carbon analyzers where the sample is acidized and introduced into a high-temperature oxygenated atmosphere to release carbon dioxide at different stages. The collected carbon dioxide at different stages is differentiated into inorganic and organic carbon. As the term indicates, organic carbon results from organic activities like the photosynthetic process by aqueous plants and from the dead remains of aquatic plants and animals. Inorganic carbon is introduced into the water through the processes of weathering.

Total carbon has implications on the acidity of the water influencing the pH and has a cascading impact on the oxygen-carrying capacity of the water. Lakes hosted by basic rocks like gabbro tend to have more inorganic carbon as compared to granite-hosted lakes. Presence of dense vegetation and reduced exposure to sunlight adds to high organic carbon in lake waters.

13.6   CONCLUSION

With increased pressure of water resources surface water is seen as the most accessible source of water for human use. Its quality, therefore, is of paramount importance in determining its use. Understanding the geological influences is therefore important in determining the type of dissolved solids expected in a water body. Water quality is not influenced by climatic and ecological conditions alone. Geology plays a very important role especially from differential weathering of various bed rocks including soils and regolith. Rock weathering and mineralogy of the host rock is critical to the water quality of surface reservoirs as evidenced by the study of two adjacent rock quarry-hosted lakes in Singapore where the quality of water was so different from each other despite being located in similar climatic and environmental settings.

Contamination of water from anthropogenic activities are increasingly impacting the present day water resources and the contaminant introduced to react with the already existing chemistry in the water which may have resulted from long processes like rock weathering and drainage accumulation from the watershed. It is therefore important to have the background understanding of how rocks underneath the water bodies are likely to have influenced the water quality of a surface water body.

KEYWORDS

- groundwater
- contamination
- water quality
- hardness of water
- rock weatherability

REFERENCES

Appelo, C.; Postma, D. *Geochemistry, Groundwater and Pollution*; 2nd ed. A. A. Balkema: Amsterdam, 2005.

Bowen, N. The Reaction Principle in Petrogenesis. *J. Geol.* **1922,** *30* (3), 177–198.

Boyd, C. *Water Quality: An Introduction*; 2nd ed.; Springer International Publishing: Switzerland, 2015.

Ceryan, S.; Tudes, S.; Ceryan, N. A New Quantitative Weathering Classification for Igneous Rocks. *Environ. Geol.* **2007,** *55* (6), 1319–1336.

Diersing, N. *Water Quality: Frequently Asked Questions*. Florida Brooks National Marine Sanctuary: Key West, FL, 2009.

Goldich, S. A Study in Rock-weathering. *J. Geol.* **1938,** *46* (1), 17–58.

Johnson, D. L.; Ambrose, S. H.; Bassett, T. J.; Bowen, M. L.; Crummey, D. E.; Isaacson, J. S.; Johnson, D. N.; Lamb, P.; Saul, M.; Winter-Nelson, A. E. Meanings of Environmental Terms. *J. Environ. Qual.* **1997,** *26*, 581–589. DOI:10.2134/jeq1997.00472425002600030002x

Ollier, C. *Weathering*; 2nd ed.; Longman: London, 1984.

Rogers, G. *United States Geological Survey Professional Paper*, Issues 117–119. Washington: Geological Survey (US), 1919, p 55.

Sarwar, S. G.; Wazir, M. A. Physico-chemical Characteristics of Freshwater Pond of Srinagar (Kashmir). *Pollut. Res.* **1991,** *10* (4), 223–227.

Zhao, J.; Broms, B.; Zhou, Y.; Choa, V. A Study of the Weathering of the Bukit Timah Granite Part A: Review, Field Observations and Geophysical Survey. *Bulletin of the International Association of Engineering Geology*, 1994; Vol. 49 (1), pp 97–106.

Index

Printed and bound by CPI Group (UK) Ltd, Croydon, CR0 4YY

23/10/2024

01777675-0011